Astronomer's Univer

Comet 2006 P1 (McNaught), as photographed by its discover, Rob McNaught, January 20, 2007.
© 2007 Robert H. McNaught

David Seargent

The Greatest Comets in History

Broom Stars and Celestial Scimitars

 Springer

David Seargent
Australia
seargent@ozemail.com.au

ISSN: 1614-659x
ISBN: 978-0-387-09512-7   e-ISBN: 978-0-387-09513-4
DOI 10.1007/978-0-387-09513-4

Library of Congress Control Number: 2008935112

Printed on acid-free paper

springer.com

For my wife Meg, and David Austen

Preface

Naked-eye comets are far from uncommon. As a rough average, one appears every 18 months or thereabouts, and it is not *very* unusual to see more than two in a single year. The record so far seems to have been 2004, with a total of five comets visible without optical aid. But 2006, 1970, and 1911 were not far behind with a total of four apiece.

Yet, the majority of these pass unnoticed by the general public. Most simply look like fuzzy stars with tails that are either faint or below the naked-eye threshold. The 'classical' comet – a bright star-like object with a long flowing tail – is a sight that graces our skies about once per decade, on average. These 'great comets' are surely among the most beautiful objects that we can see in the heavens, and it is no wonder that they created such fear in earlier times.

Just what makes a comet "great" is not easy to define. It is neither just about brightness nor only a matter of size. Some comets can sport prodigiously long tails and yet not be regarded as great. Others can become very bright, but hardly anyone other than a handful of enthusiastic astronomers will ever see them. Much depends on their separation from the Sun, the intensity of the tail, and so forth.

Probably the best definition of a great comet is simply one that would draw the attention of non-astronomers if viewed from somewhere well away from city lights and industrial haze. Typically, they are at least as bright as a reasonably bright star and sport easily visible tails, at least 5–10 degrees long. Most of the traditionally great comets of history were as bright as or brighter than a star of first- or second-magnitude with tails that could be traced to 10–20 degrees or more in a dark sky.

But these comets are not the subject of this book! What we are searching for are not simply "great" comets but the greatest of the great, the cream of the comet world. We are looking for nothing less than cometary royalty!

Picking out the best of the best is not as easy as it may sound. Ancient peoples were frequently awed by the sight of a comet in their skies, and this reaction tended to color the way they recorded it. In fact, comets were objects of such fear and dread that anything seen in the sky looking vaguely like a star with a tail was enough to trigger rumors of a comet! This can make it very difficult for a modern reader of these centuries-old records to sort out what was a real comet and what was something else.

A prime example of this is the famous (infamous?) description of a "comet" in 1528 by the French surgeon Ambroise Pare. In Pare's own words

> *So horrible was it, so terrible, so great a fright did it engender in the populace, that some died of fear, others fell sick.... This comet was the color of blood; at the summit of it was seen the shape of a bent arm holding a great sword as if about to strike. At the end of the blade there were three stars. On both sides of the rays of this comet were seen a great number of axes, knives, bloody swords, among which were a great number of hideous human faces, with beards and bristling hair.*

Very picturesque and graphic indeed! The trouble is that there are no other records of a bright comet in 1528. Whatever Pare saw in the sky, it was not a comet. Most likely, he witnessed a spectacular display of the aurora borealis. The faces, swords, and axes are probably not hard to imagine in the moving lights and curtains of a great aurora. (By the way, lest we be tempted to scoff at the naivety of our "superstitious" ancestors seeing these sorts of images in an aurora, we might recall the number of times Venus is reported today as an alien spaceship complete with landing gear and windows!).

But it is not always the original observer who causes confusion about the object recorded. A case in point is the occasional reference in modern works to a "comet" recorded by St. Augustine, probably for the year 396, that is said to have given off "a smell of sulfur." At least one book of elementary astronomy saw evidence here of the old belief in comets having an effect on the air and dismissed the reported odor accordingly.

However, what St. Augustine actually wrote was "a fiery cloud was seen in the east, small at first, then gradually as it came over the city it grew until the fire hung over the city in a terrible manner; a horrendous flame seemed to hang down, and there was a smell of sulfur." Whatever this was, it was not a comet. Augustine did not even *claim* that it was a comet. It *may* have been a meteorite fall, but an even better guess might be a lightning filled tornado funnel. The luminous effects associated with these can be very spectacular, and they are often accompanied by "a smell of sulfur"!

Incidentally, Chinese chronicles *do* record an astronomical object that year, although the description matches a nova or supernova rather than a comet. In any case, the Chinese object is almost certainly unrelated to the phenomenon noted by Augustine.

We should also be aware that, as well as dubious cases like these, some comet records are completely fictitious. A chronicler will sometimes invent a portentous comet to mark the birth and/or death of some great political personality. For example, an alleged comet appearing at the death of Charlemagne in A.D. 814 seems to have been pure embellishment.

For the most part, comets that were only mentioned on one or two nights or which appeared in a single record were eliminated straightaway as contenders for the greatest comets on record, even if their description implied something unusually spectacular. Although a minor comet might be seen on a single occasion only (and there are bona fide instances of this), anything truly spectacular is likely to have been widely observed over a considerable period of time and to have been immortalized in abundant records.

However, even after minor and dubious objects had been pruned from the list, a daunting number of entries remained. Many of these had clearly been spectacular objects that left great impressions on those who saw them. But how many could truly be listed among the *greatest* of the great comets?

For the next step in the selection, I referred to a "scale of importance" devised by D. Justin Schove in his 1984 book *Chronology of Eclipses and Comets AD 1~1000*. Although this work covered only part of the period of interest, it could be extended to earlier and later comets without too much difficulty.

Schove's scale is as follows:

1. Minor comet, noted only by experienced sky-watchers.
2. Not noted by the general public.
3. Noted by at least one contemporary chronicler.
4. Noted by some chroniclers.
5. Noted by most chroniclers.
6. Noted as remarkable in most chronicles.
7. Noted as remarkable even in short annals.
8. Created consternation. Long remembered.
9. Created terror. Remembered for generations.

After reading through Schove's list of comets and comparing them with the descriptions given in Kronk's *Cometography*, I decided as a rough rule of thumb that a scale reading of 7 or higher would qualify a comet as one of history's greatest. My aim, therefore, was to use Kronk's descriptions of Scove's 7+ comets as the standard by which to measure comets of earlier and later centuries.

In essence, this remained the method followed, although I did not always stick rigidly to Schove's evaluations and found myself disagreeing with a couple of the values he assigned.

In those cases where an orbit for the comet had been computed, and even the absolute brightness known at least approximately, it was also a helpful check to compare the comet's performance with a recent object of similar true brightness and observed under comparable circumstances. This counted as something of a reality check, especially when we are dealing with records of "frightening, prodigious signs in the sky" and so forth. Expressions like this occur mostly in early European chronicles, and they tend to conjure up images of some utterly fantastic object unlike anything seen in recent times. Yet, where an orbit allows us to form some idea of the comet's true appearance in the sky, we more often than not find it to have been something that would have fitted in very well with the brighter comets of the past few decades. Where European and Chinese records of the same object exist, the latter tend to be more sober in their descriptions and can act as another good reality check.

The end product of this pruning and sifting forms the subject of this book. What emerges is a list of over 30 individual comets, which (as far as I could ascertain from their often varied descriptions) met the criterion for being history's greatest. In addition to these comets, I have also included, separately, the historical appearances of Halley's Comet and the members of the Kreutz group of sungrazers. Among these latter are found the brightest and some of the most spectacular of all comets as well as, paradoxically, most of the smallest and faintest ever recorded.

Essentially, the comets included here were ones of exceptional brightness and/or those having long and intense tails. Yet, brightness alone or tail length alone did not automatically mean inclusion in the list. A comet might have been recorded as having a tail (say) 70 degrees long, but if there were reasons for thinking that this tail was so faint as to be missed by most casual observers, it would have been left off this list. Likewise, even comets bright enough to be seen in broad daylight were omitted if they did not also become spectacular nocturnal objects (a list of daylight comets and possible daylight comets has been added as a final chapter to compensate for what some may feel to be an unfair omission).

There will probably be objections to some of my specific omissions.

For instance, I did not include the comet of 147 B.C. The very impressive-sounding account sometimes given of this comet is, in reality, most probably a combination of three separate objects. The Chinese comet of August (for which an orbit has been computed) is not consistent with the Chinese comet of May, or with that of October and November, the latter suspected by H. H. Kritzinger as being the previous appearance of C/1858 L1 Donati. None of these objects can clearly be identified with the one recorded by Seneca sometime between the years 151 and 147 B.C. This was said to have been as large as the Sun and "so bright that it dispelled the night" – a description reading more like that of a great meteor than a great comet.

I have also omitted the comet observed by Peter Apian in 1532, despite its inclusion in most catalogs of great and remarkable comets and Apian's histori-cally important observations showing the tail as pointing away from the Sun. This seems to have been the first European recognition of this fact, although the Chinese had already noticed it as early as the ninth century.

The comet was unquestionably bright, but the tail seems to have been no longer than around 10 degrees, more in the nature of a 'typical' great comet than one of the greatest of the greats. Moreover, judging by Apian's drawings and the general descriptions of this object, its tail appears to have been predomi-nantly a plasma type. These are not as intense as the strong dust tails of very large and active comets. If Apian's Comet was rather low on dust, it is unlikely to have rated as one of history's finest, despite its obvious brightness and historical importance.

Of course, it is quite possible that I have missed some comets that should have been included, and I may have included one or two that should not be here. Moreover, there must surely have been splendid comets in ancient times that were only at their best from far southern latitudes. These would either have passed unseen and unrecorded by the chroniclers of the time or else entered into the records as relatively minor objects unworthy of being included in the present list. To these comets I offer my apologies. If recent times are any indication, some of them may even have been the greatest of them all!

In the course of the following pages, we look first and foremost at the historical returns of Halley's Comet. This is *not* because it is the biggest, brightest, and the best (it is not!) but simply because it is the most famous and the only comet that has achieved 'great' status at more than one known return. On several appearances, it has entered the ranks of the other objects in this book, having been rated variously as 7, 8 and even 9 by Schove during the first millennium of our era. Because this is a book about the greatest of the greats, the more spectacular apparitions of Halley will be the ones of chief interest to us.

Succeeding chapters will take us initially from ancient times until the end of the tenth century, then from the beginning of the eleventh until the end of the eighteenth, before moving into the more detailed modern period from the beginning of the nineteenth century to the present day.

The sixth chapter deals with that fascinating family of comets known as the Kreutz sungrazers, some of which became the most brilliant ever recorded by humankind, and the seventh with those relatively rare objects seen in daylight.

Before launching into history, however, let us take a closer look at the characters of our story – the comets themselves, what they are and from where they come.

Acknowledgments

Many people have contributed in one way or another to the genesis of this book and to all of them I extend my sincere thanks. Especially, I would like to thank Alan Garner of the Astronomical Society of New South Wales and Vic Audet of the Sutherland Astronomical Society for their records search in pursuit of an elusive article, and the following for their help in the supply of photographs and assistance in finding information from a variety of sources: Tony Marshall, Gary Kronk, Maik Meyer, Dan Green, Mary Laszlo, Joseph Marcus, Terry Lovejoy, Antoinette Beiser, Wayne Orchiston, John Bortle, R. Strom, Case Rijsdijk, Reinder Bouma, Rob McNaught, Gerard Hayes, Stuart Schneider, and David Nicholls. I would also like to thank my stepson David Austen for his invaluable help with photographic files, my wife Meg for her patience and, last but by no means least, Ms. Maury Solomon, Astronomy Editor at Springer, for her encouragement and assistance in the preparation of this book.

Contents

About the Author

David Seargent is a former lecturer in Philosophy with the Department of Community Programs at the University of Newcastle in Australia and is now a full-time writer. He is the author of the very popular *Comets: Vagabonds of Space* (Doubleday), formerly a contributing editor on comets to *Sky & Space* magazine, and currently author of the regular comet column for *Australian Sky & Telescope* (the southern hemisphere edition). He was co-author with Joseph Marcus, of a paper published in 1986 entitled "*Dust forward scatter brightness enhancement in previous apparitions of Halley's comet*" (Proceedings, 20th, ESLAB Symposium on the Exploration of Halley's Comet, Vol. 3, B. Battrick, E. J. Rolfe and R. Reinhard, eds. ESA SP-250. European Space Agency Publications). He was also the Australian co-ordinator for visual observations during the International Halley Watch, 1985–1986.

Chapter 1
The Nature of Comets

Introduction

At one time astronomers believed that the Solar System was comprised of two radically different classes of objects (actually three, if we count the single star at the center of it all). On the one hand, there were the planets and asteroids – also conveniently called minor planets, to place them in their proper planetary perspective. Although there were clear differences between the massive Jupiter and the rocky "terrestrial" planets, such as our own Earth, the similarities were great enough to ensure their inclusion in the one cosmic family. All these objects were solid and stable. The orbits they followed around the Sun were sedate, almost circular, and widely separated from one another. Asteroids, with a few errant exceptions such as Earth-crossing Apollos, shared this clockwork regularity and did not, therefore, present any great danger of rocking the astronomical boat of the Sun's planetary family.

By contrast with this well-behaved planetary family, the second component of the Solar System seemed like the proverbial prodigal son. Unlike planets and even asteroids, this second population – the comets – happily disregarded any semblance of cosmic decorum. Whereas the planetary population followed the same nearly circular orbits for eons, comets darted hither and thither in between planets and asteroids like a swarm of agitated gnats. Their orbits were anything but circular. Most of them were cigar-shaped ellipses extending from the region of the inner planets to far beyond the orbit of Pluto. Some were calculated to stretch out at least a third of the way to the nearest star. Comets in these orbits return to the planetary region only after great lapses of time.

The "periods" of many comets are calculated to reach hundreds of thousands and, in some cases, even millions of years. Orbits of others are so elongated that their period cannot even be determined with the limited data available. For these comets, the "official" orbit is simply given as a parabola, even though a perfectly parabolic orbit cannot be sustained in the real universe. In certain instances, a comet even seems to achieve the escape velocity of the Solar System, and its orbit is transformed from a very elongated ellipse into the spreading open curve of a hyperbola. These comets go off into the void of interstellar space, never to return. Or, rather, some of them do. The

D. Seargent, *The Greatest Comets in History*,
DOI 10.1007/978-0-387-09513-4_1, © Springer Science+Business Media, LLC 2009

hyperbolic orbits of others will revert to ellipses when they recede to great distances, thanks to the gravitational attraction of the Solar System as a whole. Needless to say, the final "period" of comets such as these is stupendous when compared to a human lifetime or even to the whole of recorded human history.

Not content to move in orbits as far removed as possible from those of the planets (in eccentricity if not in distance), comets are also found to pay no greater respect to the plane of the planetary system. Thus, while the Sun's planetary population orbits in pretty much the same plane – known as the ecliptic – comets have their orbits tilted each way and everywhere. A few stay close to the ecliptic plane, but most zoom in from all directions, approaching at all possible angles. There are comets that come in from below the plane at right angles and others that approach perpendicularly from above. Others approach at obtuse angles, which effectively have them moving in a direction opposite to that of the planets. These latter are known as *retrograde* orbits.

Equally un-planet-like is the appearance of comets. Instead of being stable, solid discs, comets assume nebulous, almost ghost-like, forms. Their appearance can radically change from one night to the next in a way that no planet ever would. Worse, they may even split into two or more pieces and in the most extreme cases, disintegrate altogether. That is certainly not the expected behavior of a planet!

Fig. 1.1 This view of Comet Bennett, March 27, 1970, gives a good idea of a "typical" great comet (courtesy, David Nicholls)

With these thoughts in mind, some astronomers of half a century ago felt it prudent to speak about two Solar Systems: the planetary and the cometary. Theorists such as R. A. Lyttleton even went so far as to deny a common origin for the two "systems." The planets, in the view of Lyttleton and colleagues, formed together with the Sun "in the beginning," but comets were far later acquisitions – nothing more substantial than clouds of cosmic dust clumped together and collected by the Solar System during its sporadic passages through the dark nebula that inhabit certain regions of our galaxy.

Today, the picture is at once more unified and more confusing. As astronomical discoveries began to fill in the increasingly fine details of the Solar System, the two populations became less and less distinct. Apollo asteroids went from being a handful of freaks to a populous asteroidal subsystem. Even worse, long-period asteroids in highly eccentric and steeply inclined orbits started turning up. These looked like typical asteroids but moved like typical comets! At the other end of the scale, astronomers also uncovered a population of comets moving in orbits that are more typical of asteroids!

If all of that was not enough, "transitional" objects started turning up in the lists of discoveries; apparent asteroids that sporadically sprouted comet-like tails or Earth-approaching asteroids that were found, in long-exposure images, to be enveloped in very faint veils of nebulosity.

What, then, does all this mean? What actually are comets and how do they really fit into the Solar System?

At the risk of oversimplification, we can say that a comet is actually an asteroid largely made of ice – nothing more, nothing less. Think of the minor, sub-planetary members of the Solar System as being arranged on a sort of spectrum with hard, dry, and rocky or rocky–metallic bodies at one extreme and fluffy conglomerations of ice and dust mixed together (as it turns out) with organic tar on the other. Those on the "dry" end are asteroids and those on the volatile end are fully active comets. In between lies a variety of ice-rock bodies that either spend most of their days as inert asteroids, with occasional bouts of cometary activity, or as weakly active comets amounting to little more than asteroids occasionally surrounded by thin and extremely extended "atmospheres."

Although we will look a little more closely at the differences between comets and the broad types of orbits these objects follow, let us just note at the moment that comets on the more or less "asteroidal" end of the spectrum are usually of short period (although there are exceptions), whereas those having very long periods and nearly parabolic orbits appear to be quite fragile, icy bodies.

So, in the end it may be best think of a comet as an icy asteroid. Please do not, however, form the mental picture of a white and pristine snowball! The ice is far from pure. For one thing, comets contain not only water ice but also a mixture of various frozen gases. In addition, there is a meteoric dust component as well as a rich variety of quite complex organic compounds. Perhaps a better description of a comet would be a mass of frozen mud – or even a mass of frozen muddy froth, considering the low density of much cometary material. One thing is for sure: you would not be adding lumps of cometary ice to your cocktails!

Nucleus of Comet Halley
P.J. Stooke, 1996

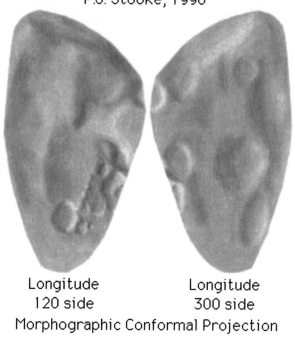

Longitude   Longitude
120 side   300 side
Morphographic Conformal Projection

Fig. 1.2 A map of the solid nucleus of Halley's Comet drawn by Phil Stooke, Department of Geography, University of Western Ontario, from data obtained by spacecraft during the 1986 return (courtesy, NASA)

As this mass of low-density icy mud approaches the Sun, the latter's warmth causes the surface ices to boil off into surrounding space. In the vacuum of space, the melting point and the boiling point of water are one and the same, and water ice behaves in the same way as the "dry ices" or the frozen gases with which it is mixed. Water vapor and other gases boil out of the frozen body and into the surrounding void. As they do, particles of the "mud" are also released and carried away from the solid body to join the ever-expanding cloud that has begun to surround it. Solar radiation excites atoms of gas and causes them to glow by fluorescence, rather like Earth's polar lights or the gaseous contents of a neon sign. The particles of mud (which we will more correctly call as "dust" from now on) reflect and scatter incident sunlight. Both the ice and dust contribute to making the cloud visible in our telescopes.

It is this hazy cloud that we see as the coma of a comet. The word "coma," by the way, means *hair* in this context and is so named because of its typical fuzzy appearance. It has no association with a prolonged period of unconsciousness!

The coma of a comet is an immense object, in dimension if not in mass. Some very large comets have sported comas having diameters greater than the Sun itself, although quarter to half a million kilometers is more typical.

By comparison, the central icy asteroid – technically known as the "nucleus" of the comet – is a tiny thing. There are some giant ones measuring tens of kilometers or even larger, and there are also Lilliputian ones less than 10 m (approximately 33 feet) across, but the majority of nuclei are found within the 1–10 km (roughly, 0.6–6.3 mile) range. These smaller ones, however, are most often far from spherical in shape, and their length is often considerably greater than their width. Some of those observed close up from space probes have been likened to sweet potatoes. But whatever their shape, it is remarkable that from objects such as these – bodies that would sit comfortably within the perimeter of a moderately sized city – the great nebulous comas are formed, clouds that dwarf the biggest planets and that occasionally swell to sizes larger than the Sun itself!

It is one of the paradoxes of comets that these small and fragile objects can not only generate such huge comas but that they are capable of doing it again and again, even after repeated close encounters with the Sun. Halley's Comet, to use a famous example, possessed a coma over a million kilometers (625,000 miles) across during its 1986 return. Yet, it has been generating comas of this dimension for thousands of years, each time sweeping past the Sun within the orbit of the planet Venus! We might think that something so fragile and icy would have broken up and vanished long ago.

The reason why a comet such as Halley's can go on producing comas return after return is the very tenuous nature of the coma. Although enormous, comet comas contain relatively little matter. By the standards of Earth's atmosphere at sea level, the coma is a hard vacuum. Halley's, and similar comets, lose the equivalent of a couple of meters girth each time they pass through the inner Solar System. For a body 10 km in diameter, this is not a very great shrinkage!

Nevertheless, although one return (or even 100 returns) may make little difference to a comet, eventually the nucleus will be exhausted and the comet will disappear. The only alternative to eventual disintegration is a close approach to Jupiter, resulting in the comet being kicked into a different orbit that keeps it well clear of the Sun, or the development of an insulating crust of dusty material on the nucleus' surface blanketing underlying ices from the Sun's heat. A third alternative, collision with one of the planets, offers a more violent (and rare) means of disintegration!

Actually, all of these alternatives from disintegration to planetary collision have been observed or inferred. The famous Shoemaker-Levy 9 hit on Jupiter was a spectacular example of the latter, but the demise of comets through breakup and disintegration has also been observed, and there is good evidence that some comets have been damped down into asteroids, presumably through the growth of an insulating crust. Asteroids in comet-like orbits may truly be old comets that can no longer produce comas, and there is evidence that some comets go through long periods of dormancy, periods that may eventually stretch out into permanent extinction. A prime suspect is the short-period comet Denning–Fujikawa. This comet was discovered in 1881 by Denning and, despite a period of just 9 years, was not seen again until its rediscovery by Fujikawa in 1978. Both times it was a relatively bright and active object visible in small telescopes. Except for rather fast-fading and a probable sharp rise and fall in brightness, it appeared pretty normal for a short-period comet. But why had it not been seen between the two discoveries? What is more to the point, why has it not been seen since, even though the 1987 return should have been very favorable and potential observers had the advantage over its discoverers of knowing where to look for it?

It seems that this comet spends most of its time as a very faint and dormant asteroid, presumably crusted over with an insulating layer that keeps its ices from direct exposure to sunlight. Once in a while, we may imagine a piece of the insulation breaks loose for some reason or other (perhaps meteorite impact or thermal or tidal cracking) and Denning–Fujikawa bursts briefly into full-blown cometary activity. We might call this the "Brigadoon comet," only coming to life, like the fabled Irish village, once every 100 years!

Earlier, we mentioned that the light by which we see the coma is a combination of fluorescing gases and sunlight reflecting and scattering off particles of dust. The contribution from these two light sources is not always the same but varies from one comet to the next and may even vary for the same comet at different times in the same apparition.

When the light of some comets is passed through a spectroscope, it is found to consist almost entirely of the emission lines of various molecular species. In most instances, the visual region of the spectrum is dominated by the three so-called Swan bands of diatomic carbon. In these comets, the solar continuum of light reflected from the dust component is very weak and confined to the brighter central regions of the coma, very close to the central nucleus.

By contrast, other comets are so rich in dust that the solar continuum dominates, effectively swamping the gas emission lines in the visual spectrum. In these comets, not only is there a bright continuum in the nucleus region, but the coma itself and even the tail can, in the most extreme cases, be devoid of gaseous emissions.

It would really be more accurate to speak about two comas, the gas coma and the dust coma, and for the study of the dynamics within a comet, this is a distinction that must be made. However, as the two occupy (more or less) the same region of space, there is no need for us to be so pedantic here.

There is, however, a third component of the coma that we should distinguish, namely, the neutral hydrogen coma. If we think that the visual coma is big, this third component becomes almost unbelievable. But because it radiates only at ultraviolet wavelengths, it remained completely unknown until the advent of space-based observatories. It was discovered in early 1970 by the first orbiting UV observatory (OAO – Orbiting Astronomical Observatory) in UV images of the comet Tago-Sato-Kasaka. This comet – not an especially large one – was found to be surrounded by a tenuous cloud of hydrogen one million kilometers in diameter. A couple of months later, the great comet Bennett was shown to have an even larger hydrogen coma, and in more recent years, Comet Hale-Bopp of 1997 was found to possess a hydrogen cloud some 150 million kilometers (93 million miles) across. The diameter of this hydrogen coma was equal to Earth's distance from the Sun!

The hydrogen for these vast clouds is supplied by photo-disassociation of water vapor molecules within the visible coma. Once again, we marvel at the paradox of a small object such as a comet nucleus giving rise to something that even on a cosmic scale is large, exceeded only by the largest supergiant stars, galactic nebula, and entire stellar systems.

When the coma of a typical comet is viewed through a small telescope, its appearance is much like an unresolved globular star cluster or, as an unidentified friend of the famous practical astronomer Sir Patrick Moore is reputed to have said, "like a small lump of cotton wool." Unless the comet is only weakly active, there is normally a marked brightening toward the center. When this is present, the comet is said to be "centrally condensed." There is actually a 0–9 point scale of degrees of condensation that comet observers use in their visual descriptions. A degree of condensation (DC) of zero means that the coma is totally diffuse, with no perceptible central brightening, whereas a DC of nine means that the comet appears either as a stellar point of light or a small planetary disc with little or no trace of diffuse coma.

However, as if to make matters a little more confusing, the term "central condensation" can also refer to a discrete feature within the central regions of the coma. Certain comets are not only centrally condensed in the sense of brightening steeply toward the center of their comas, but also display a central "core" that may appear either as a small fuzzy disc or else as an almost star-like point. Either way, it stands out as being more or less differentiated from the general concentration of brightness at the coma's center.

A comet may, however, be described as being centrally condensed without having a true central condensation in this sense. When the central condensation (in the sense of a discrete core) is very intense and bright, it is very often referred to as a "false nucleus," "photometric nucleus," or (simply and unfortunately) "nucleus." This last is technically incorrect and very confusing. Very rarely is the true physical nucleus – the solid, icy, asteroidal body from which issues the phenomena that make comets what they are – discernible. In some comets, a discrete central condensation and a definite photometric nucleus are both discernible. In these comets, the central condensation typically appears as a small central disc in low-power eyepieces and appears to be the core of the comet. However, when carefully examined through a powerful eyepiece, an even smaller "core within the core" is visible, normally as nothing more than a faint star-like point of light. Technically speaking, this, not the larger and more conspicuous central condensation, should then be termed the photometric nucleus.

When a comet is active, the feeble light of the nucleus – that is, the true, physical nucleus – is swamped by the far brighter glow of the inner coma and photometric nucleus. Unless a comet passing close to Earth shows only the weakest activity and sports a coma that is nothing more than a gossamer thin veil, the best chance of observing a true comet nucleus is after the comet has receded far from the Sun and its activity has all but shut down. Large telescopes armed with CCDs may then detect it as a very faint speck of reflected sunlight. What is sometimes referred to as the nucleus of an active comet in older literature is simply the central condensation, or maybe the photometric nucleus within the *central region* of the central condensation. Either way, the term applies to a region far more voluminous than the solid body itself. Early estimates of the nuclei of comets that gave values of hundreds or even thousands of kilometers certainly did not refer to the true nucleus. Clearly, they were measures of the far larger central condensation.

Comet observers like to see comets displaying a sharp central condensation. Other things being equal, these comets tend to be active objects and, if they are moving along orbits that will bring them close to the Sun, can become visually impressive. Although not an iron-clad guarantee, a bright and sharply defined central condensation is welcomed as a positive sign. A comet showing a sharp condensation early in its apparition (i.e., while still relatively far from the Sun) and which is also destined to venture within Earth's orbit, holds a good chance of developing an impressive pseudo-parabolic coma with the central condensation brightening into an intense false nucleus at its focus. The outer comas of comets such as these form relatively distinct envelopes rather than the indistinct boundary of the more typical globular coma. Sometimes, there is even a series of concentric envelopes and jet-like structures emanating from the photometric nucleus. Most impressive of all, though, these are the comets that traditionally develop the best examples of the phenomenon that has come to characterize comets in the popular mind, namely, the "tail." We will now turn to this spectacular feature.

The Tails of Comets

Ask the average non-astronomer what word comes to mind when comet is mentioned, and the answer will most likely be either "Halley" or tail! Yet, the majority of comets observed nowadays actually display very little tail when observed visually. Photography and CCD imaging do far better at detecting tails, but the typical appearance of a faint comet in the eyepiece of a modest telescope is that of a fuzzy coma with, at best, a minor extension in a direction away from the Sun. The grand appendages that have struck such wonder and terror in the collective psyche of humanity since time immemorial are not typical of comet tails in general.

Another popular misconception by the person in the street is that comet tails relate to the speed of the object. The spectacular tails of great comets do indeed mimic trails left in the wake of a speeding body, but in reality that appearance is nothing more than an illusion. In the near perfect vacuum of outer space, there is insufficient resistance to sweep material back into a train of this type.

Although it might superficially resemble a wake left by something speeding by, a comet tail actually has a very different genesis.

As the Chinese have known for over a 1000 years and the Europeans since Peter Apian's observations of the Great Comet of 1532, comet tails basically point away from the Sun, irrespective of the direction of motion of the comet itself. This implies that when a comet is moving inward toward the Sun, it is moving coma first (we can now safely say "head first," as the coma/central condensation/nucleus is termed the "head" when a tail is visible). However, when a comet is moving outward from the Sun, it goes away tail first!

Clearly, something emanating from the Sun pushes material away from a comet and into the tail. For a long while this repulsive force was thought to be sunlight but, although partially correct, the real situation is more complex than this. A better understanding of the process requires us to distinguish two basically different types of tails.

Recall that earlier we mentioned two types of coma – gas and dust – but passed it by as a needless complication for our purposes. Well, the same distinction carries over to comet tails where, however, it assumes too great an importance to be casually set aside!

The broadest division of comet tails is, therefore, into gas (or more accurately "ion" or "plasma," as the gas is ionized in these features) and dust. Tails of the first variety are known as Type I and the second (predictably) as Type II and Type III. The difference between Types II and III is minor and can be overlooked for the present. (There are also rare and only recently detected sodium and iron tails, but as these are not visually discernible they need not concern us here.)

Type I tails are traditionally straight, tend to be long, and when well-developed consist of a bundle of narrow thread-like rays diverging from the central region of the coma. Small and weak tails of this type are much less impressive, most often appearing as nothing more than a single faint ray emerging from the center of a coma. Comets whose tails are predominantly of this type also tend to have globular comas.

Well-developed and active Type I tails make for very spectacular images, but are unfortunately a lot less impressive when viewed by eye. Being streams of ionized gases, they radiate principally in the blue region of the spectrum, to which the human eye is not particularly sensitive. Unless they are especially intense, we tend to see Type I tails as being rather dim and disappointing.

Type I tails are directed almost precisely away from the Sun. As the comet swings around the Sun, tails of this type show little distortion but sweep around like a searchlight beam as they maintain their relatively strict anti-solar orientation.

On the other hand, Type I tails may (apparently without warning) experience the most fantastic contortions. They have been seen on occasion to develop warps and kinks of up to 90 degrees. At other times, comets have shed their tails altogether, only to immediately sprout new ones in their place. The old tail, or the portion of it that was set adrift, takes the form of an elongated cloud, unconnected with the comet as it drifts away. These peculiar happenings are known as disconnection events or DE's.

Following an idea originally proposed by S. Arrhenius and developed in the first decade of the twentieth century by K. Schwarzchild and P. Debye, it was thought that the pressure of sunlight alone, acting upon the ions in the gaseous coma, was responsible for the occurrence of Type I tails. Certainly, light does exercise pressure, as we will see shortly, but astronomers came to doubt its ability to explain the motion of discrete tail features such as fast-moving knots and kinks. Once the velocities attained by some of these were measured with tolerable accuracy, it became clear that something else was involved. Although Debye had shown that radiation pressure from sunlight could account for forces of repulsion that exceeded solar gravitational attraction by a factor of 20 or 30, studies of the motions within the plasma tails of very active comets such as Morehouse of 1908 and Whipple-Fedtke-Tevzadze of 1943 indicated repulsive forces exceeding solar attraction by as much as 100–200 times, quite beyond the capabilities of sunlight. Even worse, to account for the very narrow thread-like streamers so often photographed in Type I tails, repulsive forces as great or greater than 1,000 times that of solar attraction were required! Clearly, something else was being emitted by the Sun; something that exercised a far greater repulsive force on particles of ionized gas than sunlight alone could accomplish.

As long ago as 1893, J. M. Schaeberle proposed that material particles ejected from the Sun were the cause of comet tails. Writing in the *Astronomical Journal* he proposed that, "The tail of a comet is produced by the visible particles of matter originally forming the comet's atmosphere, and by the previously invisible particles of a coronal stream that, moving with great velocity, finally produce by repeated impact of the successive particles almost the same motions in the visible atmosphere of the comet as would be communicated by a continuously accelerating force directed away from the sun."

Although Schaeberle's idea did not catch on at that time, it was remarkably close to the truth. Type I tails, we now know, are formed by the so-called solar wind, which (as Schaeberle opined) boils outward from the solar corona. Essentially, it forms an extension of the corona itself. It is this "wind" of protons and free electrons, streaming outward at velocities of 1,000–2,000 km (625–1,250 miles) per second, that carries away cometary ions into those long streaming tails of plasma. Turbulence within this wind reflects in turbulent motions within the tails, and even such dramatic and seemingly unpredictable events as tail disconnection events can be explained in terms of magnetic polarity reversals within the solar wind. For this reason, Type I tails have been graphically described as "solar windsocks," and prior to the advent of artificial satellites capable of directly measuring the wind, they were the only way of monitoring this phenomenon.

By contrast, light pressure alone appears quite adequate to account for the more sedate motions within dust tails.

Tails of this variety, though usually not very detailed photographically, tend to be more visually apparent than those of plasma. This is because we see them by means of sunlight reflected off, and scattered by, myriads of fine dust particles. Our eyes are more sensitive to this continuum spectrum of sunlight than to the blue glow of Type I tails, even if photography and other imaging techniques are not!

The more leisurely pace of dust tail particles is betrayed by the morphology of these appendages. For one thing, although they also extend away from the general direction of the Sun, they do not stay as close to the strictly anti-solar vector as Type I tails do. As the dust particles travel further from the comet's head, they depart more and more from this strictly anti-solar direction. Moreover, not all the particles of dust in a Type II tail have identical masses. Less massive particles experience a greater degree of acceleration, by light pressure, than that experienced by larger ones, and they will therefore be accelerated away from the comet at greater velocities than their larger companions.

The trajectories of small dust particles will, therefore, lie closer to the strictly anti-solar direction than those of the more massive ones. Consequently, the paths of the latter are more strongly curved as they increasingly lag behind the anti-solar vector.

The overall result of this divergence of dust-particle trajectories is a delicately curving tail, widening away from the head as the differing degrees of curvature become more apparent. These curving features do not reach full development, however, until after the comet has passed the Sun. On the inward leg of its orbit, a comet (if displaying a dust tail at all) will normally possess a relatively straight and fairly short appendage, even though this may at times be spectacularly bright.

Fully developed Type II tails, despite their sometimes considerable length and breadth, are quite thin. Dust particles swept back from a comet's head have very little tendency to drift either above or below the plane of the comet's orbit, causing the tail to be remarkably flat within the orbital plane. If Earth is located such that we see the comet from a perspective more or less perpendicular to its orbital plane, we will view the dust tail face on. If the tail is an intense one containing plenty of dust, it will be spectacular and assume the famous scimitar shape of a classical bright and dusty comet. On the other hand, if the comet has not been shedding a great deal of fine dust – if it is either a gassy comet or one whose dust is too coarse to feed strong Type II tail development – this feature may be so faint and transparent as to pass unnoticed.

Nevertheless, if or when Earth moves through the comet's orbital plane and we are in a position to see the tail edge-on, the dust tail will quite suddenly emerge from obscurity, though not as a classical curving scimitar. From a perspective within the plane of the comet's orbit, we view the tail edge-on, peering through its entire width. Needless to say, its curvature is not apparent, and the tail will appear to us as a long beam of light having approximately the same width as the coma. Because line-of-sight curvature is not apparent, we could gain the mistaken impression that the Type I tail had inexplicably intensified ... except that the characteristic tail rays found in these appendages remain absent.

An excellent example of this effect was exhibited by Comet Austin in 1990.

Discovered on December 6, 1989, by Rod Austin in New Zealand, this comet was at first visible only from southern latitudes, slowly brightening as it approached the Sun and drifted into evening twilight. Passing just 0.35 AU from the Sun on April 9, 1990, the comet emerged into the Northern Hemisphere pre-dawn sky as a faint naked-eye object sporting an impressive ion tail visible through binoculars.

Because the comet approached Earth even as it receded from the Sun, it remained a dim naked-eye object for an extended period. Nevertheless, its appearance changed significantly during that time, with the ion tail fading away and the coma ballooning out from a small and compressed head into a large, diffuse ball of nebulosity. For many visual observers, the comet had become a bland and tail-less globule. There were a few reports of a tail, but this was so faint that most observers failed to detect it.

During the weeks following the comet's re-emergence after its encounter with the Sun, we were viewing it more broadside than edgewise with respect to its orbital plane. Austin's was not a dusty comet, so this observational perspective meant that any dust tail that may have formed would have been faint and very difficult to detect.

During early June, however, Earth passed through the plane of the comet's orbit, and for an interval of several days the dust tail emerged from obscurity as a beam of light streaming for at least 3 degrees away from the globular head, almost as bright and intense as the coma itself. One might have supposed this to be a rejuvenated ion tail, except that it lacked the typical plasma tail ray structure. Clearly, the Type II tail of this rather dust-poor comet had been too faint for general observation, except for the brief period that it presented edge-on to our planet.

Of course, the same type of enhancement takes place for bright dust tails, and the result is then a very intense tail, "like the beam of a powerful searchlight" to quote one graphic description, actually an account of Halley's Comet in May 1910 when its dust tail turned edge-on to our line of sight.

Earlier, we mentioned another form of dust tail; the so-called Type III tails. These are defined by dust particles so large that the pressure of sunlight has only a small effect upon their trajectories. Seen face on, these tails are characteristically short and very strongly curved, but when viewed from certain angles, and especially from within the plane of a comet's orbit, the effect of perspective may actually project the tail in front of the comet such that it appears to point *toward* the Sun. This is referred to as an anomalous tail, or anti-tail, though "pseudo-anomalous" would probably be a better term, as no tail material actually exists between the comet and the Sun in these instances. When viewed at or very near the time of Earth's passage through the comet's orbital plane, these tails take the form of a narrow spike directly opposite the main tail, as was spectacularly displayed by Comet Arend-Roland in 1957. Because the particles comprising them are large and sluggish, tails of this type are unlikely to be present until after a comet has rounded the Sun, as only then will the coarse dust have had time to spread far enough from the comet to form the feature.

Genuine anomalous tails – true anti-tails – do exist, but they are even rarer than the pseudo form. The anomalous tails displayed by Comets Seki-Lines in April 1962 and McNaught in January 2007 were seen at times when Earth was well away from these comets' respective orbital planes and really did involve the presence of heavy dust particles on the sunward side of their orbits. The anti-tails of these two comets appeared more as extended sheaths of light than as the familiar sharp spikes of the Arend-Roland type.

Features Within Dust Tails

We earlier described dust tails as being rather featureless in comparison with those of Type I tails. But we should also note that there have been some spectacular exceptions to this, as can be seen just by glancing through some of the photographs in this book!

When dust tail features are present, they normally take the form either of broad and more or less diffuse streamers diverging from the central regions of the coma and extending in the direction of the tail itself, or of multiple striae running more or less across the breadth of the tail and seeming to converge toward a point located "in front" of the nucleus, i.e., between the head of the comet and the Sun. Not that they actually extend to this point in space, of course, but if we were to extend imaginary lines through them, these would converge somewhere on the sunward side of the comet, in a region free of any cometary material. The actual features themselves seem to originate along the inner concave edge of a curving Type II tail, almost as if they are radiating out from the focus of this curve. The inner edge of these tails, incidentally, is usually more diffuse and less sharply defined than the outer, convex, edge, and it appears that the striae somehow originate within this diffuse tail boundary.

Dust-tail features of the first type, which we have simply called "streamers," are quite distinct from the sharp and narrow rays of the ion tail. They are quite readily explained, but before we can appreciate this explanation, we must first introduce two terms that relate to the process of dust-tail formation, "syndynes" and "synchrones."

Without needlessly digressing into the long story of dust-tail dynamics and the pioneering work of F. W. Bessel in the first half of the nineteenth century and Th Bredikhin in the second, we may simply note that these terms denote two families of curves followed by particles in the dust tail.

Dust-tail particles, as already mentioned, come in a wide range of sizes. As we also remarked, the size of a particle makes a big difference in its trajectory as the pressure of sunlight pushes it away from the comet's head. The path that any particle takes through space is determined by the counteracting forces of the Sun's gravity and the pressure of sunlight. The former attracts the particle toward the Sun as the latter repels it, and the actual path taken by the particle can be looked upon as a sort of compromise between these two opposing forces.

The more massive the particle, the less effect the repulsive force will exercise on its trajectory. The less massive the particle, the weaker will be gravity's influence on its path. At any moment, the entire range of particle sizes and masses will be smoking out of the nucleus of an active comet. What syndyne/synchrone analysis does is chart (a) the paths taken by particles of different sizes released at some specific instant of time and (b) paths followed by particles having the same size released over an extended period of time. These matching particles will experience the same degree of repulsion by radiation pressure and will therefore move away from the nucleus at matching velocities. The curve given by (a) is called a synchrone and that by (b) a syndyne (or, syndyname in Bredikhin's terminology).

Perhaps a more homely simile will give a better appreciation of this analysis. Imagine a cross-country foot race in which the contestants are of varying degrees of fitness. There are some professional runners, but there are others who might, perhaps, have been better employed watching from the sidelines.

The starter's pistol fires, and all the runners surge forward en masse. Very quickly, however, the different levels of fitness start to make their mark. As the professional runners surge ahead, so the less accomplished progressively lag behind. Those with some degree of aptitude for long races are behind the leaders but not too far. Those with little physical stamina soon find themselves well behind the leaders of the field.

After a while, we end up with a line of runners strewn out across the field, extending backwards from the most capable runners and ending with the least. If the race is a long one, the distance between the leaders and the end of the line might be great, even though all the runners set out at precisely the same instant, and all entered the event as a compact grouping. The resultant line of runners is analogous to a synchrone.

For a similarly sporting analog of a syndyne, we may imagine another group of runners who are not racing against one another so much as against the clock. They are competing in an endurance run from point A to point B in the shortest possible time. Imagine, further, that they did not begin their run at the same instant, but that their starting times were spread over a period. Moreover, imagine that they are not all permitted to take the same route. The more able runners are handicapped by being required to run along a steeper path, whereas progressively slower runners are directed toward increasingly easier routes. An observer looking over this field of runners from a suitably elevated position could therefore determine the different abilities of the various runners, not by noting how quickly they covered the distance from A to B or how close they were to either point but by noticing which routes the different runners followed. Those on the steeper slopes are the fittest and fastest, those on the gentlest slopes, the least fit and the slowest.

The similarity between these fanciful athletic events and the adventures of particles emitted into a comet's tail are not difficult to appreciate. As a range of athletic abilities was found among our field of runners, so a cloud of dust emitted by a comet nucleus at any specified instant comprises a wide diversity of particle sizes. Like runners at the discharge of a starter's pistol, all of these particles leave the nucleus together, but are quickly and progressively sorted out by their differing responses to the repulsive effect of solar radiation. Large particles will acquire only a modest acceleration in the anti-solar direction and will soon start falling behind the field of smaller particles for which the push of sunlight has a far more powerful effect. A puff of dust from an eruption on the nucleus surface will assume the form of a line of dust as particles of differing sizes are sorted out, much as our group of runners soon stretched itself into a line. In the dust tails of comets, these lines are observed as the dusty streamers of which we earlier spoke. In more technical language, these features are *synchrones* within the dust tail.

On the other hand, dust particles of similar size will experience a similar degree of acceleration from sunlight and will therefore follow one another along the same paths in the tail irrespective of when they actually emerged from the nucleus. Of course, their time of release will determine how far along the tail they have progressed at any specific moment, but the actual track followed remains constant with particles of similar size following one another through the same region of tail. These paths are *syndynes*, and it is the range of these that determines the morphology of a dust tail. Just as the observers of the cross-country endurance run could tell which runners were more and which were less accomplished simply by noting the paths they followed, so can astronomers determine which regions of a dust tail are comprised of large particles and which of small by calculating the syndynes predicted for particles of varying sizes and fitting them to the actual shape of the tail.

Syndynes of large particles occupy the inner, concave part of a curving dust tail, while the more sharply defined leading (convex) edge coincides with the predicted syndynes of the smallest tail particles. Quite simply, these latter are the syndynes that come closest to being straight; they represent the trajectories of tiny dust motes swept away at velocities most closely approaching – though always, of course, remaining less than – the accelerated ions of plasma tails.

Syndynes and synchrones nicely explain the overall shape of dust tails, as well as accounting in a straightforward way for the existence of the rare dust streamers.

But what about the slightly more common (albeit still rare) striae of which we spoke earlier and which have been noted in the dust tails of several bright comets? Can the model account for these as easily?

The second bright comet of 1957, Mrkos, provided a fine example of these features, and an analysis by William Liller of Alan McClure's spectacular photographs of this object concluded that they were indeed "terminal synchrones" explicable in terms of the Bessel–Bredikhin model. This conclusion was quickly challenged, however, by the often controversial Lithuanian comet expert S. K. Vsekhsvyatskij. Vsekhsvyatskij argued that the mechanical theory of Bessel and Bredikhin failed for three reasons.

First, striae always seem to occur in a definite sequence, all having approximately the same length and breadth and all separated by approximately equal distances. They also seem to be of about the same intensity. All of this is difficult to explain if these structures are the products of random eruptions on the nucleus.

Secondly, Vsekhsvyatskij was also worried by the fact that, if extended, these streaks would converge at a point between the nucleus and the Sun. This should not occur if their origin was in the nucleus itself.

Thirdly, the structures are always of short duration, typically lasting only a few days, rarely much longer than a week.

Vsekhsvyatskij made a good case against the striae being synchrones, but he tended to overreact by going on to cast doubt on the very idea that these features were even comprised of dust! As an alternative, he proposed that the material making up the striae is a mix of heavy polyatomic ionized molecules that align themselves along magnetic fields. According to Vsekhsvyatskij's proposal, the patterns of striae are, in effect, "magnetic maps," not unlike those made in all elementary science classes by sprinkling iron filings over a sheet of paper held above a bar magnet!

This proposal had serious problems of its own, however. It was not easy to see how these heavy ions could remain hidden in the pure continuous spectrum of a dust tail or, for that matter, how they came to be there in the first place. In consequence, this proposal won few converts.

On the other hand, at least some of Vsekhsvyatskij's colleagues gave more credence to his criticisms of the mechanical model. K. Wurm, to name just one, although not at all convinced of his heavy ion idea, nevertheless thought that the introduction of magnetic fields was a positive insight. Wurm proposed that, through exposure to solar radiation, dust particles should lose electrons by the photoelectric effect and become electrically charged and therefore susceptible to magnetic fields. This, he suggested, might explain the striae according to Vsekhsvyatskij's magnetic model while at the same time avoiding the need to introduce unobserved and difficult to explain heavy ionized molecules.

Wurm's proposal seems a good one in so far as it introduces no ad hoc processes. Both magnetic fields in interplanetary space and the photoelectric effect are known, and the striae do *look* as though they are aligned along magnetic lines of force. Nevertheless, this hypothesis never managed to gain much support. (Interestingly, though, a very similar mechanism has been proposed to explain the spokes in Saturn's rings, discovered by space probes in recent years!)

The explanation of striae that is nowadays most widely accepted is the one put forward by Zdenek Sekanina and developed in his analysis of the complex dust-tail structures exhibited by Comet West, the Great Comet of 1976. This comet was unusual in that it displayed *both* striae and true synchronous streamers. The latter quite obviously originated in the nucleus region and diverged through the tail in a fan-like formation. Striae, on the other hand, originated along the inner boundary of the tail. At times, striae and rays met and formed angles of around 10 degrees to one another.

Noting that the tail region from which the striae appeared to originate was also that occupied by large particles, Sekanina argued that the former were products of myriad disruptions of these relatively massive particles. The large particles themselves were released from the comet's nucleus, of course, but after they had traveled some distance from the head, thermal and/or dynamical stresses caused them to rupture (perhaps progressively) into fine dust which, being more susceptible to acceleration by sunlight pressure, was swept along a different trajectory from that of the parent particles. In this way, both the intensity of the striae (fine dust giving an effectively greater reflecting area than an equal mass of coarse particles) and the apparent diffusion of the streaks from a point on the sunward side of the nucleus are explained. The more or less simultaneous disruption of a stream of large and fragile particles would also appear capable of explaining the short duration of the pattern of striae, as well as their relatively regular appearance.

By the way, it is interesting to note the similarity of appearance between these striae and the plumes of fine mist formed within the spray curtain of a water jet. Although there are many dissimilarities, of course, it is difficult not to think of a striated dust tail as a fountain of particles breaking into an even finer spray curtain!

If fine dust does indeed constitute striae, we would (other things being equal) expect the section of dust tails exhibiting these formations to be brighter than the featureless regions. This is dramatically supported by the tail of Comet Ikeya-Seki, the Great Comet of 1965, in which the section of tail containing striae was noticeably more intense than the adjacent section on the sunward side; this is just the opposite of what we would normally expect!

The Brightness of Comets

An asteroid or inert comet nucleus, shining only by reflected sunlight, varies its true or intrinsic brightness according to the inverse-square of its distance from the Sun, modified only by the phase effect. Neglecting this latter for the present, it follows that if the asteroid's distance from the Sun is halved, its intrinsic brightness will increase by a factor of four. Similarly, if its distance from the Sun remained constant such that its apparent brightness (i.e., how bright it appears to us) depended solely on its distance from Earth, it would brighten and fade according to the inverse-square of its distance from Earth

For an asteroid orbiting the Sun, the apparent brightness depends upon a combination of its distance from the Sun and from Earth, plus the modification introduced by the effect of phase.

This brightness, like that of any astronomical object, is given in terms of the scale of stellar magnitude. This is a scale of brightness in which an object of a given magnitude is approximately 2.51 times brighter than one of the next magnitude, with the smaller figures denoting the greater brightness. For instance, an object of second magnitude is 2.51 times brighter than an object of third magnitude and is, in turn 2.51 times fainter than an object of first magnitude; an object of magnitude –1 is 6.3 times brighter than one of +1 and so forth.

A difference of 5 magnitudes corresponds to a difference of 100 in brightness and one of 10 magnitudes, to around 10,000, while the Sun at about magnitude −26.7 is some 300,000 times brighter than the (approximately) −13 magnitude full Moon. The faintest star visible by naked-eye near the zenith from a rural location is around magnitude 7, although magnitude 6 is considered to be the naked-eye limit under most "good" conditions.

The stellar magnitude of an asteroid can be calculated by

$$m = H_o + 5 \log \Delta + 2.5n \log r + \text{phase}$$

where m is the apparent magnitude, H_o is the absolute magnitude (i.e., the magnitude that the object would have at 1 astronomical unit (AU) from both Earth and Sun; an AU being equal to Earth's distance from the Sun, in round figures, 150,000,000 km) and r and Δ are the distances of the asteroid from Sun and Earth, respectively, each given in AU. The parameter n can be thought of an index of the sensitivity of the object's brightness to its distance from the Sun. As we are dealing with something shining merely by reflected light in this instance, n will equal 2 (i.e., the inverse-square of the object's distance from the Sun). The brightness formula for an asteroid can, therefore, be simplified to

$$m = H_o + 5 \log \Delta + 5 \log r + \text{phase effect}$$

Now, an active comet is a different matter entirely. The true physical nucleus behaves like an asteroid, but its feeble contribution to the total light of a comet is, in most instances, so meager as to be completely swamped by the glowing gases and reflected/scattered sunlight of the gas and dust comas. The proportion of light from fluorescing gas molecules and that reflected off, and scattered by, dust particles differs from comet to comet and even for the same comet at different times during its apparition. For some comets, as we have already remarked, most of the light comes from glowing gas while very dust-rich ones are visible almost entirely by reflected sunlight.

Because fluorescing gases and reflecting dust clouds differ in important ways from inert solid objects such as asteroids, we should not be surprised that the inverse-square law no longer holds; the parameter n rarely equals 2 in active comets. In fact, an early study of 45 individual comets by Bobrovnikoff (in the early 1940s) found values of n ranging from 11.4 to −1.77. In terms of the above formula, this translates as a range from 28.5 $\log r$ to −4.43 $\log r$. Very large values tend to be transitory and values less than 2 (especially negative values) can be pathological, indicating disintegration of the comet. But even if we take these rare extremes out of the picture, values of n anywhere from about 2.5 to 6 or even 8 would not raise too many eyebrows amongst comet observers.

Naively, we might suppose that the higher values of n should be found among gaseous comets and that those objects whose light is chiefly the solar continuum reflected off dust particles will more nearly approach the $n = 2$ value of asteroids. However, this is not necessarily true. Some gaseous comets have large values of n

and some dusty comets have very small ones, but the converse has also proved true at times. The rate at which a newly discovered comet will brighten is not settled simply by determining whether it is dust-rich or dust-poor!

A complicating factor is that comets are active, dynamic bodies. As an icy nucleus warms on approach to the Sun, both gas and dust are expelled in varying quantities. If the number of dust particles within the coma of a very dusty comet remained constant, it would vary in brightness according to an n value close to 2, assuming that the light from glowing gases contributed only negligibly and neglecting any phase effects that would come into play under certain observing geometries. In actual fact, though, the number of dust particles will not remain constant. As the comet ventures closer to the Sun and activity increases on the nucleus surface, we would expect (other things being equal) the number of dust particles being released to increase proportionally to the decline in the comet–Sun distance. As the comet draws closer to the Sun, not only will it catch and reflect more sunlight, but the effective reflecting surface area will also increase as more dust particles are added to the coma. The disruption of large particles into smaller ones probably plays a role as well, and the overall effect can be a value of n well in excess of 2, even though most of the light continues to come from reflection.

Estimates of the average value of n in comets have varied over the years. A major study of comets from the earliest recorded times until the middle of the twentieth century was carried out by Vsekhsvyatskij, who concluded from this study that most comets follow an n value very close to 4 (i.e., 10 log r in the formula). A value of this order seemed to make sense as it fitted quite nicely with the expected behavior of a fluorescing gas cloud. It was consequently used for many years as the default value in predictions of a newly discovered comet's expected brightness development. Even today, it is frequently employed where a specific value for n has not been determined, and it also continues to serve as the basis for a standard of absolute magnitude comparisons between comets.

As we stated earlier, the absolute magnitude (H_o) of a Solar System object is the brightness that it would have if observed at 1 AU from both Sun and Earth. In this way, it is possible to make a direct comparison between the real or intrinsic brightness of a sample of objects. For inert asteroids – or major planets, for that matter – this is fairly straightforward, as the value of n is already set at 2, and it is a simple matter to compute their absolute from their apparent brightness. But for comets, the value of n could be almost anything and because the value of H_o depends critically upon this, the usefulness of the latter as a means of comparison is compromised. If an average value of n were to be found, however, the H_o of comets could be reduced on the assumption of average brightness behavior. This parameter would then be a much more useful standard of comparison.

Because of the huge influence of Vsekhsvyatskij's research, his average of $n = 4$ became widely adopted. Indeed, comets continue to be compared with one another according to their absolute magnitudes computed on the assumption that $n = 4$. This form of absolute magnitude is denoted by the symbol H_{10} and is still thought of as a sort of "canonical" value for the comparison of comets.

Nevertheless, even in his own day Vsekhsvyatskij's conclusion did not go unchallenged. His chief critic was N. T. Bobrovnikoff, whose research in the early 1940s has already been mentioned. As we saw in our above reference to this study, a very wide range of n values was noted. Nevertheless, Bobrovnikoff was able to derive an average for his sample of $n = 3.2$ (8 log r). His sample was a lot less extensive than Vsekhsvyatskij's, however, and for that reason his average was not used as often in predicting the brightness of comets.

Other early studies suggested different values of n for comets having different orbital periods. Thus, in their study published in 1951, J. Oort and M. Schmidt derived average values of n for old comets (those whose orbits suggested many previous returns) of 4.2, in rough agreement with Vsekhsvyatskij, but only 2.8 for comets that appeared to be making their first approach to the Sun. We will look a little later at this issue of old and new comets, but for the present let us just say that the orbits of some comets suggest that they have never passed through the inner Solar System before and are, at least in the dynamical sense, new objects.

The following year V. Vanysek published his results of a somewhat more comprehensive study involving 99 comets observed from 1853 to 1951. He found that comets of very short period (the majority having periods of less than a decade) had average values of $n = 4.9$, those with periods in the range 125–11,000 years had average n values of 4.6, for those with periods from 11,000 years to near parabolic, $n = 3.3$, and for those whose orbits were so elongated as to be indistinguishable from parabolas, $n = 3.1$. Most, though probably not all, of these latter would be new in the sense of Oort and Schmidt. Unless the orbit of a newly discovered comet was quickly found to be an obvious short-period ellipse, chances are it would have a period somewhere between the low hundreds of years and the high hundreds of thousands and so a value of $n = 4$ might not seem too far off the track unless an obvious departure from this trend became quickly apparent. For an obviously short-period object, a default value of $n = 6$ came to be widely used (although Vanysek's results implied that $n = 5$ would have been better).

About 20 years after Vanysek's study, David Meisel and Charles Morris tackled the problem afresh using a more comprehensive data base of 150 comets of all orbital categories. They also took account of certain systematic errors that crept into brightness data due to some of the methods that had been used to determine the visual magnitude of comets and that might have corrupted earlier statistics to some degree. These authors found that for all the comets in their sample (which ranged from short period to new), n averaged a value of 3.6. For the old comets (a couple of which were periodic) the average was $n = 4.2$ and for the new ones, 3.2. There was, however, a considerable scatter of data points in all of these categories.

In more recent years (principally since 1990 or thereabouts) a default value of $n = 3$ (7.5 log r) has been increasingly used in the International Astronomical Union's *Circulars* for predicting the brightness behavior of newly discovered comets, although the more extensive predictions published in the *Minor Planet Circulars* generally retain the traditional $n = 4$ default value.

It is also recognized today that many (probably most) comets do not have a fixed value of n that remains constant throughout their periods of visibility. Typically, the rate of brightness increase slows as a comet nears the Sun, and more complex formulas reflecting this are sometimes employed in brightness predictions and light-curve analyses.

The tendency for comets to exhibit larger values of n when further from the Sun can at times be nothing more than an observational artifact. When the outer coma has a very low surface brightness – as it may before a comet fully activates – it can escape detection, either by instrumental or visual means. What is observed is only the more intense region of the innermost coma. Although this small region is more intense, its contribution to the total light of the coma is relatively minimal. It is not unusual for the estimated brightness of this region to be three or four magnitudes fainter than the entire coma, and if that coma is so diffuse as to be consistently missed, brightness estimates of the comet will be faint by that amount!

However, as the comet's activity increases and the surface brightness of the extended coma intensifies, a point will be reached where the latter crosses the threshold of visibility. An observer following the comet's progress will find that it quite suddenly seems to expand and surge in total brightness. For a brief period, the value of n appears to go through the proverbial roof – estimates of 8, 10, or even higher being possible!

Nevertheless, there is also a genuine tendency for some comets (especially those approaching the Sun for the first time) to display genuinely larger values of n when they first become active relatively far from the Sun. Orbiting our star at great distances for eons of time while perpetually frozen at temperatures only slightly warmer than absolute zero, a comet nucleus is widely thought to acquire a layer of highly volatile material on its surface. It is probable that mild warming from radioactive decay within the nucleus causes volatiles such as carbon monoxide to migrate upward and freeze at the nucleus surface. Cosmic ray bombardment over billions of years tends to promote the accumulation of very unstable damaged molecules within this surface layer as well. So long as the comet stays clear of the Sun, this frosting will remain inert, but if it is deflected by a passing star or interstellar dust cloud into a new orbit taking it to within the distances of the inner planets, the gentle warming of the approaching Sun begins to have a dramatic effect.

At distances comparable with those of Saturn, the frosting layer becomes active and the comet develops a very premature coma and, not infrequently, tail. These early features are believed to be comprised, for the most part, of ice particles rather than the more refractory dust shed by comets closer to the Sun. With its activity driven by very volatile and unstable substances, the comet may, for a time, brighten steeply. Strictly speaking, this behavior could be called anomalous, a sort of prolonged and very early brightness outburst that can give a totally erroneous picture of what the comet's true H_o and n might be.

As the comet drifts closer to the Sun, the frosting layer boils away, exposing the less volatile ices beneath. In addition, the ice particles of the coma and tail evaporate at these higher temperatures; the tail shrinks and the comet passes into a more sedate phase of activity. If it relied solely upon the very volatile layer of frosting, a comet like this would shut down altogether, but with decrease in solar distance, the main nucleus constituent of water ice (as hard and stable as basalt at the distance the comet first sprung into activity) starts to feel enough warmth to enter its own active phase. Effectively, the source driving the comet's activity switches by degrees from very volatile ices to the less volatile water ice, the transition making its presence apparent in a significant decrease in the rate at which the comet is seen to brighten and therefore in the numerical value of n.

On occasions, this drop can be as sudden as it is extreme and has been the cause of some pretty embarrassing incidents. There are notorious cases where astronomers have gone on record predicting that a newly found comet will evolve into a spectacular sight, only to end up wishing that they had kept their forecasts to themselves! A hypothetical example will show how this may happen.

Suppose a 16th magnitude comet has been discovered at a distance of 6 AU from the Sun and 5 AU from Earth and that early brightness estimates suggest it to be brightening according to $n = 8$. On that value, the absolute magnitude is computed to be a whopping -3! Now, imagine further that the orbit of this comet reveals that it will approach the Sun to just 0.3 AU, at which time it will also be 1 AU from Earth. Although no astronomer in his or her right mind would actually assume such a thing, if the comet's n value were to hold all the way in, its brightness would increase to a fantastic -13.5, about equal to the full Moon!!

Now, suppose that the value of n dropped to just 2.8 soon after discovery. Far from having $H_o = -3$, the comet would then settle down to a modest absolute magnitude of just 7. Peak brightness near the Sun would be downgraded to 3.3, still naked-eye, but certainly nothing like the full Moon!

This hypothetical example is, of course, deliberately extreme, but it serves to illustrate the point. It also explains why most cometary astronomers have become very cautious about going public with predictions these days, assuming that if one is going to err, it is best to err on the side of conservatism.

The Motions of Comets

The way comets move across our skies was for a long time a mystery to humankind. The fixed stars maintained their regular patterns, faithfully marking out the times and the seasons. Even the wandering planets at least kept to the regular path of the ecliptic and maintained a consistency about their wanderings. Comets, on the other hand, seemed the ultimate cosmic anarchists. They appeared in any region of the sky. Even the circumpolar constellations where no planet ventured were not immune to the traverse of comets. Moreover, the comets seemed totally oblivious to the times and seasons, and their motions across the sky were anything but regular. Some moved so slowly that they spent their entire period of visibility in the same region of the heavens. Others, by contrast, fairly raced through the skies, traversing several constellations during the course of only 1 or 2 days. Yet others appeared to hover in the same region for a period, then race across tens of degrees of sky in a couple of days, only to stagnate again at a spot far from that of their first appearance. For centuries, such behavior seemed radically at odds with the regularities of the celestial vault.

Not surprisingly, therefore, many early thinkers saw in comets a meteorological rather than an astronomical phenomenon. The Mesopotamians were said to have understood comets to be "a kind of eddy of violently rotating air," at least in the opinion of the Greek Epigenes. The hypothesis that became most widely accepted, however, was that proposed by the great Greek philosopher and pioneer scientist, Aristotle (384–322 B.C.). According to this famous thinker, comets were exhalations from Earth (broadly speaking, eruptions of gases, probably volcanic in origin) that, upon reaching the fiery upper regions of the atmosphere, ignited. Sometimes the exhalations will burn furiously and quickly, and we see a meteor or shooting star. At other times, the fire is slow and may burn for days or even weeks. These slow-burning fires are the ones which, in Aristotle's view, we observe as comets.

Although about as far from the truth as it could possibly be, this model is actually supported by observational evidence! Bright comets with long tails tend to be located near the horizon, with their tails directed upward into the sky. Although we now realize that this simply means that most are close to the Sun at the times of their most conspicuous display, it takes little imagination to picture them as gaseous eruptions from the ground, bursting into flame upon reaching the upper atmosphere.

Some form of this meteorological model held sway for centuries. Significantly, Ptolemy's *Almagest*, the greatest astronomical treatise of ancient times and one that determined the cosmological thought of humanity for over a thousand years, did not even mention comets.

In later centuries, other explanations were put forward from time to time. Following Tycho Brahe's demonstration that the Great Comet of 1577 showed no discernible parallax, comets could no longer be thought of as nearby phenomena. Tycho's parallax experiment placed the 1577 comet at least several times further than the Moon, unambiguously placing it within the cosmic realm.

Not all scientists of the day believed that comets were astronomical *objects,* however. For instance, Galileo, although apparently agreeing that they were located beyond the terrestrial realm, denied that they were material objects at all. In one of his few serious mistakes, Galileo argued that comets were a form of optical illusion, not unlike rainbows, induced by reflections of sunlight. No less than Ptolemy, Galileo denied that the study of comets lay within the province of the astronomer.

Other ideas were starting to come to the fore, however. Thus, Johannes Kepler (1571–1630), despite giving comets an astrological significance, put forward the idea that they were accumulations of impurities with tails comprised of "filth" forced out by the Sun's energy. Despite his rather inelegant choice of words, this account of comet tails was a remarkably accurate insight. His idea of their orbits was not as insightful, though. He held the view that comets move through space in straight lines, in this respect differing from his predecessor, Tycho, who preferred circular orbits. Johannes Hevelius (1611–1687) saw comets as fragments that had broken away from Earth, the Sun, and the planets and believed them to move through space along parabolic trajectories.

This issue of the paths of comets grew in importance once they came to be accepted as bona fide astronomical objects and not terrestrial gas eruptions or cosmic rainbows. But it was Newton's discovery of the phenomenon of gravity that finally enabled the matter to be settled. It was Newton who finally tamed the wild comet.

Not a cosmic anarchist at all, a comet could for the first time be shown to obey the same laws of motion as the stars and planets. Newton's discovery showed that these objects moved through space along a curve that was a conic section with the Sun at one focus. The only real difference between the orbit of a comet and that of a planet such as Earth lay in its eccentricity; whereas Earth follows an elliptical orbit that departs only slightly from a circle, most comets follow orbits that are almost parabolic.

The sequel to Newton's work is well known. His colleague, Edmond Halley (1656–1742), calculated the orbits of a number of comets and, although the positional measurements available to him only allowed parabolic orbits to be calculated, he came to suspect that at least some of them were in fact very elongated ellipses.

Halley was especially intrigued by the striking similarities between three of the objects in his catalog. The comet that he had observed in 1682 was found to move in an orbit remarkably similar to that seen by Kepler in 1607 and to another described by Apian in 1531. So similar were the orbits of these comets that Halley made the revolutionary suggestion that all three were actually different appearances of the self-same object. He proposed that this comet orbited the Sun in an ellipse so elongated as to pass the Sun's vicinity (where alone, it became bright enough to be seen from Earth) at intervals of 75–76 years. Such an elongated ellipse would be indistinguishable from a parabola in the data available to Halley.

As further support for his idea, a bright comet was also noted in the year 1456, and, although Halley had not computed the orbit of this object, its position in the sky and general motion was consistent with it being an even earlier return of the comet of 1682.

On this evidence, Halley made the bold prediction that the comet would again re-appear around the year 1758. The rest, as they say, is history. As we will see in Chapter 2, the comet was once again sighted on Christmas night, 1758, and became a bright naked-eye object early the following year. It has since returned in 1835, 1910, and most recently in 1986. It is due back again in 2061.

Although Halley did not live to see his prediction verified, he was honored by having his name for ever after associated with this, the most famous of comets.

If three of Halley's comets were really a single object returning on three occasions, maybe all the apparently parabolic orbits in his catalog were really elongated ellipses, and all comets return to our region of space sooner or later. Maybe the chief difference between the other objects in his study and the 1531/1602/1682 object lay in the comparative brevity of the latter's orbital period, decades instead of centuries, millennia, or longer.

Actually, Halley did suspect the existence of at least two further periodic comets. He suspected that the comet seen following the assassination of Julius Caesar in the year 44 B.C., one recorded in a Byzantine chronicle in 530 A.D., and the great daylight comets of 1106 and 1680 (the very one whose orbit was studied extensively by Newton in his *Principia*) were actually separate returns of the same object. This speculation, as it turns out, was not right. These comets truly are separate and unrelated objects. Indeed, the Byzantine comet of 530 A.D. turned out to be none other than Halley's Comet itself!

Halley also suspected the comets of 1532 and 1661 to be related, probably the same object on two separate returns. This again has proven to be incorrect, although not before the story was given an extra twist by the return of the second comet in 2002! Now known as Ikeya-Zhang, after its 2002 discovers Kaoru Ikeya and Daqing Zhang, this comet is probably also identical with those recorded in 1273 and 877 A.D. – but not with the comet of 1532. Nevertheless, the published orbit of this comet is so similar that until the period of Ikeya-Zhang was determined, it was widely suspected as being a return of the 1532 comet. If the computed orbit of this earlier object is relatively accurate, there may still be an association of some sort. It is possible that both comets are major fragments of an object that broke apart hundreds of years ago and that the 1532 comet might also return at some date in the not-very-distant future, or maybe even have returned already under very unfavorable circumstances and escaped detection.

In general, though, Halley's insights were correct. Most comets do orbit the Sun in highly elongated ellipses and take centuries, millennia, or hundreds of millennia to make a single circuit. Being so close to the parabolic limit, the gravitational pull of one of the larger planets such as Jupiter is sometimes enough to boost the comet into a weakly hyperbolic orbit and send it out of the Solar System forever. On the other hand, no comet has ever been observed moving in the type of strongly hyperbolic orbit indicative of an origin elsewhere in the galaxy. Though they may become far flung, comets are true members of the Sun's family.

When a comet is discovered, the first priority is to calculate an orbit with sufficient accuracy to prevent it from being lost. Needless to say, neither is the initial orbit very accurate nor does it need to be. As long as it predicts the comet's position for the next several nights with tolerable accuracy, that is enough. Further observations will enable astronomers to improve the orbital computations and extend the predicted path. For the sake of simplicity, the

initial orbit is almost always calculated on the assumption of it being a parabola. Strictly speaking, this assumption is wrong 100% of the time, but it makes
for easier math, and the chances are high that the real orbit will be so close to a
parabola that the difference will only become apparent after the comet has been
tracked across a significant arc of sky. In fact, the difference might be so close as
never to be discovered, and the published orbit may never appear as anything
other than a parabola.

The typical comet orbit has the shape of an elongated cigar. The point closest
to the Sun (where the comet turns around and heads for home) is called the
perihelion point and that furthest from the Sun, where the comet starts again to
fall back toward the inner Solar System, is called its *aphelion*. The degree of
elongation of the orbit is known as its *eccentricity*, a perfect circle having an
eccentricity of 0, a perfect parabola 1, and the spreading open curve of an
hyperbola >1. A comet's period is the time (normally given in years) taken for it
to make one full revolution of its orbit, best thought of as the time between
successive perihelion passages or returns of the comet.

There is no need to wait for a comet to return to find its period. It can be
calculated simply enough from the formula,

$$P = (q/(1 - e))^{3/2}$$

where P is the period in years, q is the perihelion distance in AU, and e is the
eccentricity.

Of course, to derive very accurate periods, astronomers need to do more than
simply use this formula. The gravitational perturbations of planets must be
taken into account as well as (whenever possible) the thrust effect of gases
escaping from the nucleus itself. Both of these effects influence the eccentricity
and therefore the period of a comet.

The plane of a comet's orbit itself is, unlike that of the planets, not necessarily
close to the ecliptic. Comets may approach the Sun from any direction, above or
below, the ecliptic plane. This is why they show no preference for the zodiacal
region of the celestial dome.

A subsystem of comets exists with aphelia near Jupiter's orbit. These objects
with periods from 5 to 7 years are known as short-period comets. It is believed
that they have been recruited through a long series of planetary perturbations
from a belt of icy objects beyond the orbit of Neptune known as the Kuiper Belt.
Larger members of the belt are being discovered regularly today and (despite
howls of protest from some quarters) that enigmatic object known as Pluto and
long listed as the ninth planet, has now been officially recognized as a member
of this population. The members of the central family of short-period comets
can therefore be thought of as little brothers and sisters of Pluto!

Comets of longer period, however, originate far beyond the Kuiper Belt. In
1950, Jan Oort compared the orbits of 21 long-period comets for which accurate
orbits were known, and discovered a peculiar tendency. Instead of a random

scatter of aphelia, there appeared to be a disproportionate concentration at distances between 50,000 and 100,000 AU from the Sun. This tendency was confirmed by a second study in 1953, in which he included a total of 41 comets of very long period. More recent statistics with ever widening databases have increasingly supported Oort's early findings.

These comets were coming in from such vast distances as to be almost on the border of escape from the Solar System. Their average period was around 40 million years, some much longer! So tenuous was the Sun's hold on these objects that the gravitational disturbances to their motion inevitably experienced during their passage through the planetary system would drastically alter their orbits. Oort calculated that about half of the comets in his sample would be ejected from the Sun's grasp on hyperbolic orbits. The other half would have their orbital eccentricities seriously reduced, in most cases reducing their aphelion distances to the order of 10,000 AU and their periods to only 400,000 years or thereabouts. Oort and his colleague Schmidt reasoned that, because it would be freakish for a comet to maintain the 40 million year type of orbit for more than one complete circuit, those comets found in such orbits must be on their maiden voyage into the Sun's planetary system. It is possible that a minority of these objects was ejected into very long orbits from a smaller ellipse at a previous return. However, their total numbers were just too great for this to be the chief explanation. It seemed that only two alternatives were available. Either there was a continuous creation of comets at distances of between 50,000 and 100,000 AU or a storehouse of comets existed out in these regions – an extensive reservoir which the Estonian astronomer Ernst Opik (who independently reached a similar conclusion to Oort's) described as "a vast sphere occupied by the comets." Oort and Schmidt opted for the latter alternative, as they saw no known mechanism capable of forming comets at these distances.

We should mention briefly that a proposal for a variety of cometary continuous creation *was* put forward about the same time by British cosmologist R. A. Lyttleton. Lyttleton proposed that comets formed through the accretion of cosmic dust via the gravitational lens process during the Solar System's passage through clouds of interstellar material. According to his theory, however, comets would form at intermediate distances from the Sun, a lot closer than the aphelia of the objects in Oort's study. Moreover, the Lyttleton scenario predicted a totally unsatisfactory comet model. Lyttleton rejected the entire notion of a solid nucleus in favor of a model in which comets were merely clouds of dust particles and each particle followed its own orbit around the Sun. These individual particles were supposed to converge into a visible cloud close to their common perihelia. Explanations were given for the appearance of photometric nuclei and the development of tails, but the model has now been totally disproved by direct observations of the physical nuclei both from the ground and outer space.

As we remarked earlier, the orbits of comets with very long periods show no respect for the plane of the planetary orbits. The cloud of comets must therefore be pretty much spherical in shape and can be pictured surrounding the disc-shaped planetary system as an analog of the spherical halo of globular star clusters surrounding the disc of the Milky Way Galaxy.

Oort's comets with the longest periods, those running into millions of years, must truly be new, not in the sense that they were formed only recently on the cosmic time scale but in the sense that until relatively recently they have remained in cold storage at the very outermost perimeter of the Solar System. Although the question of origins is still being debated, there is a growing consensus that these objects were ejected from the planetary regions during the time of the Solar System's formation. There they remain, orbiting the Sun along vast ellipses that carry them no closer to the planetary regions.

Once in a while, however, a passing star or interstellar molecular cloud will disturb some of these remote comets, setting them on a long fall inward, toward the distant Sun. Although the chance of falling all the way into the Sun is very slight, some of these will come close enough to activate and, maybe, be found by terrestrial astronomers. A few found their way into Oort's catalog!

Following Oort's discovery, astronomers normally refer to these comets as being "new in the Oort–Schmidt sense" or simply as dynamically new in order to differentiate them from those comets that are simply new in the sense of having been recently discovered.

From the frequency of discoveries of dynamically new comets and estimates as to how many were probably missed, Oort determined the approximate influx of these objects into the inner Solar System and, by further estimating the chance that a comet within the "sphere of comets" (which quickly became known as the "Oort Cloud") will be deflected to within the range of our telescopes, he estimated the total population of the outer regions to be in the order of 100,000 million. Some more recent estimates have multiplied this by a factor of ten or even a hundred, although more conservative estimates have also been made. Whatever the true number, though, it is vast and far, far exceeds the number of active comets recorded throughout the whole of human history.

A puzzling factor that Oort noted in his statistics was the dramatic drop in the number of comets as we move to smaller aphelion distances. We recall that it was this that first drew his attention to the existence of the remote comet sphere, but it is nevertheless odd that the one half of the population of dynamically new comets that Oort estimated to return along orbits of shorter period and less remote aphelia should not have shown up more clearly in his statistics. Where is the greater portion of the dynamically new comets that passed by 400,000 years ago and which should now be returning along their smaller orbits?

Oort proposed that most dynamically new comets either disintegrate or fade dramatically following just a single perihelion passage within planetary distances of the Sun. Indeed, dynamically new comets have acquired a reputation for unpredictable behavior. We have already discussed their tendency to brighten more slowly than older or more dynamically evolved comets, and it is also true that they have a greater tendency to experience rapid fade-outs, breakup, and other disruptions. Yet, it is also true that the majority do not obviously disrupt or disintegrate after passing by the Sun, although, of course, we can say nothing of their *long-term* stability after passing from our sight. It is just possible that their period of activity close to the Sun triggers a slow but progressive process of disruption that might continue for years, centuries, or even millennia after the comet has faded from view. We simply do not know one way or the other.

In Oort's view, the dynamically old comets that we see returning on orbits of (comparatively) shorter period constitute a more durable tail of the population. Further planetary influences will tend to shorten the orbits of many of these still further until the most enduring of the durable tail end up in relatively short-period orbits of the Halley type. Very few, though, will wind up as very short-period comets. These, as we have already said, are now believed to originate within the closer Kuiper Belt.

There is, however, another theory as to why there are comparatively few second-time-around comets. Oort assumed that the influx of dynamically new comets remains constant over time, i.e., what we see today is the steady-state influx that has existed throughout most of the life of the Solar System. But suppose if some 20 million years ago the Solar System passed through or close to a cluster of stars or a giant molecular cloud and an unusually large number of Oort Cloud comets were deflected into the planetary region, they would be arriving within the inner planetary system right now. Suppose, in other words, that throughout the recorded human history, Solar System has been experiencing a *comet shower*.

Comet showers of this nature might last for millions of years, but if the present one (if, indeed, there *is* a present one) began sometime within the last 400,000 years or so, the first few comets of the shower may not have had time to return on their shorter orbits. The relatively low rate of returning comets might actually reflect, not the fragile nature of most dynamically new comets, but the true steady-state (non-shower) rate of these objects' arrival within the planetary system. Maybe at some time in the distant future (say about 100,000 years from now), dynamically new comets will be comparatively rare while "second timers" returning on orbits of around 400,000 years or less will be commonplace. There may even be times in Earth's history when a new influx of Oort Cloud comets making their debut in the inner Solar System and those from an earlier comet shower making their encore performances overlap. These would be times when comets arrive in abundance within the planetary region. Maybe, during these epochs devastating collisions with Earth will occur, leaving their signature in mass extinctions like those found in our planet's fossil record.

It is also interesting to speculate that if our era is indeed experiencing an unusual influx of first-time comets, this has given us a remarkably good opportunity to learn about the Oort Cloud. If the bunching together of aphelia at great distances – the very thing that gave astronomers their main clue to the cloud's existence in the first place – really does result from a temporary comet shower, it will disappear once the shower ends. Had our era been one of steady-state influx, the aphelia of very long-period comets would have been more evenly distributed.

Without the distant bunching that so impressed Jan Oort, astronomers would probably not have become aware of the presence of the Oort Cloud!

Of course, this would have been a major handicap to our knowledge of the origin of comets, but it may have had wider ramifications as well. For example, many astronomers now believe that the stability of Oort Clouds around other solar systems play a large role in the habitability of these systems. For instance, a solar system that passes frequently through the spiral arms of its home galaxy must suffer frequent disruptions of its own Oort Cloud and experience frequent and intense comet showers. For an Earth-like planet in such a system, it would be like living at the target end of a cosmic rifle range! Ignorance of the existence of Oort Clouds might make such a planet look a lot friendlier to life than it really is.

To delve any further into such matters would, however, take us far beyond the scope of this book. Accordingly, we must now leave our necessarily brief introduction to comets in general and pursue our primary quarry, those comets which we deem the greatest of the greats, the "Olympic champions" of the comet world.

We begin with the comet everybody has heard about. It has not always rated as one of the greatest of the greats – some of its appearances have been less spectacular than others – but it stands alone as being the only comet of relatively short period that regularly becomes a very striking naked-eye spectacle.

This comet can be, of course, none other than Halley, the very worthy subject of Chapter 2.

Chapter 2
Halley's Comet Through the Ages

Ask anyone picked at random out of a crowd to name a comet, and we could bet that the name given would be "Halley." The reason is not hard to find. With a period of around 76 years, a single trip around the Sun is so close to the Biblical "three score and ten" of human life expectancy that it has gained the title of "the once in a lifetime comet," and the memory of its apparitions get passed down from parent to child like a cosmic family heirloom.

Because it is so widely known, some urban myths have grown up around it. Some folk have actually been surprised to learn that there are other named comets (though perhaps not as much in this post-Hale–Bopp, post-McNaught era), and for a long time there was a belief that Halley's was the most spectacular of all the comets. This was probably influenced by accounts of the 1910 appearance (at times, dare we say, viewed through the magnifying glass of memory) and the fact that most of the really bright comets between 1910 and 1957 favored the southern hemisphere and were therefore missed by most of the world's population. This urban myth has no doubt been largely corrected by a comparatively dim 1986 apparition of Halley coupled with some very spectacular objects in the decades immediately preceding and immediately following. Still, Halley's remains *the* comet in the minds of many people.

Earlier, we mentioned D. Justin Schove's list of comets seen during the first millennium and his estimation of these on a nine-point scale. We suggested that, other things being equal, those with scores of seven or higher should be included in this present book as the greatest of the great comets of history. In Schove's list, the comet later identified as Halley's scored seven for the returns of 66, 374, 530, 607, 684, 760, and 912; eight for 837; and a whopping nine in 141!

Schove's scoring is principally based on how widely observed a comet became and how long its appearance remained in the collective memory of a community. Normally, an outstanding comet would induce greater terror and remain in the memory longer than a mediocre one. Reading the actual records of Halley's at various apparitions during the first millennium, however, does not give the impression that it deserved the high scores awarded by Schove. It was certainly bright and would have been spectacular, but its appearance alone should not have made it so widely recorded as to deserve scores as high as seven or better on most of these returns. From the recorded descriptions alone, it is

D. Seargent, *The Greatest Comets in History*,
DOI 10.1007/978-0-387-09513-4_2, © Springer Science+Business Media, LLC 2009

doubtful if it should have rated as one of the greatest of the greats at returns other than those of 141, 374, 607, and 837, during the period covered by Schove and at returns other than 1066 and 1910 at later times. By the way, the larger number of greater displays during the first millennium is not due to an intrinsic fading of the comet but to small changes in the orbit during the past 1,000 years or so. In the first millennium, the comet could pass closer to Earth than it can today, and the spectacular returns of 374, 607, and 837 all saw approaches closer than 0.1 AU to our planet. The 141 return was not quite as close (0.17 AU), but the comet was very well placed in the sky at that time. Actually, despite Schove's rating and the very brief and matter-of-fact account of the 374 return that has come down to us, this apparition was almost certainly more spectacular than that of 141 as the comet is known to have passed closer to Earth that year than at any other return except that of 837. At the time of closest passage, though, it was better placed in the southern hemisphere, and its full grandeur passed unrecorded for posterity.

The returns of 1066 and 1910 were moderately close (0.10 and 0.15 AU, respectively), but on each occasion the comet passed through favorable forward scatter geometry and became unusually bright.

Forward scattering of sunlight is a phenomenon that occurs when small particles of matter are present between Earth and Sun. Homely examples include the "silver lining" at the edges of clouds obscuring the Sun and the brilliant illumination of drifting spider's web and thistle down passing before the Sun on a clear day. We cannot see drifting web, for example, until it passes almost in front of the Sun, when it suddenly emerges from obscurity as brilliant threads of silver. Dust particles in comets experience the same phenomenon, and when a dusty comet is observed at large phase angle (that is to say, more or less between Earth and the Sun) it becomes anomalously brilliant. As we shall see later in this book, several comets have become visible in broad daylight when very close to the Sun principally due to the forward scattering effect. In 1066 and 1910, Halley's appears to have joined their ranks! Moreover, during these two returns, the dust tail was seen edge on, making it seem unusually long and abnormally intense.

But if the magnitude and tail length of Halley's Comet cannot normally account for this object's observation record in ancient times, what is the reason for its observability in days before astronomers and the general public knew in advance of its approach?

The reason has more to do with the orbit than with the comet itself. Most comets, irrespective of their periods, are only seen to advantage during relatively small windows of opportunity. There is a restricted range of dates during which perihelion passage can fall if the comet is to be favorably placed for observation from Earth. Too far outside that range, the comet will be poorly placed and may not even be visible. Take Comet Hyakutake (the Great Comet of 1996), for example. The chief reason why this comet became "Great" was its approach to just 0.1 AU of Earth at the time it was very favorably placed high in the skies and well away from both the evening and morning twilight. But for such a close approach to happen, the comet's perihelion needed to fall very close to the actual date on which it occurred in 1996. Otherwise, Hyakutake would not have passed near Earth and would have appeared no more impressive than many other relatively bright comets of recent years. Chances are that at the comet's previous return, back in pre-historic times, it was less spectacular, and when it again comes back thousands of years hence, it will most probably be less impressive also.

But that is not true for Halley's Comet. Here, the orbits of both comet and Earth are related in such a way as to allow two broad "windows" of observation. While most comets, if they can pass close to Earth at all, have only one narrow opportunity on either the outward or the inward leg of their orbits, Halley's can make a close approach on the inward (i.e., pre-perihelic) leg and on the outward (or post-perihelic) leg. These days, the very closest approaches are restricted to the latter when, fortuitously, the comet is also intrinsically brighter and more active.

When perihelion falls early in the year (northern spring/ southern fall), the closest approach to Earth occurs after perihelion. Just how close this will be depends on the actual time of perihelion passage, but at least a relatively close approach can occur for a wide range of perihelion dates. For many of these dates, the comet will also be well placed in the sky at the time of closest passage of our planet. The southern hemisphere is favored at these returns. The closest approaches possible nowadays would see the comet come just within 0.1 AU of Earth and be located almost at the south celestial pole. Such an event will actually happen in 2134, when Halley will pass just 0.09 AU from Earth and 8 degrees from the South Pole on May 7.

If perihelion occurs in the northern fall/early winter or southern spring/early summer, relatively close passages of Earth will occur some weeks prior to perihelion. Once again, of course, the exact details depend upon the actual date of perihelion passage, but in general these apparitions favor the northern hemisphere and we can see the comet (as in 1378) passing very close to the north celestial pole. That year saw the comet a mere 0.12 AU away and just under 7.5 degrees from the pole on October 2. The comet is normally a little less active and a little fainter, intrinsically, before perihelion than after, and the tail typically less dusty then. These "late-year" returns are therefore not quite as spectacular as the "early year" ones, although as seen against a dark sky the comet's plasma tail can be very impressive and typically spans 15–30 degrees or thereabouts. Moreover, being very well placed for the northern hemisphere, these apparitions had the advantage in earlier times of being seen by more people and more adequately recorded.

Returns around the time of New Year are the least favorable, as the comet approaches and recedes more or less from behind the Sun. Nevertheless, unlike most comets, this "window of non-opportunity" is narrow, and the last time Halley is computed to have come to perihelion around this date is back in 690 B.C., long before its first appearance in historic records.

Among the least favorable historical returns are those having perihelion passage mid-year. The problem with these is that the closest approach to Earth occurs close to perihelion, and, because perihelion is itself a little more than half an AU from the Sun, this necessarily means that the comet can approach Earth no closer than a little *less* than half an AU. Moreover, the comet does not ride as high in the dark sky as at most returns, but its relatively small distance from both Sun and Earth somewhat compensates with a good peak in brightness. In fact, these relatively poor apparitions of Halley are not dissimilar to the favorable apparitions of great comets such as Donati in 1858 and Bennett in 1970! Examples of mid-year returns include those of 451, 1456, and the predicted next apparition of 2061.

This is one of the chief reasons why Halley's Comet has enjoyed such an unprecedented history of observation. In terms of observing geometry, even a relatively poor return of this comet would be considered a favorable apparition for most other objects, and its intrinsic brightness is enough to ensure that, even on these returns, it can hold its own, if not with the *very* greatest of comets, at least with the majority of those that have still been designated as great over the centuries. As for its favorable returns, Halley rides high in the sky of one or other hemisphere and provides the rare spectacle of a bright comet far from the twilight zones. Once again, unless the apparition is an exceptionally favorable one, it may not be the greatest of comets, but it is still bright and (in part, just because of its location in a dark sky) very conspicuous. Even when only mediocre great (so to speak) it can still attract the sort of attention that would normally be reserved for the grandest of comets.

The story of Edmond Halley, his comet, and its importance in the history of cometary astronomy has already been touched upon in this book and has been told and re-told so many times that many readers will probably know it backwards. However, no account of Halley's Comet would be complete without a few words about Halley and his discovery.

Edmond Halley (1656–1742) was born the son of a wealthy London soap-boiler, became a Fellow of the Royal Society in 1678, Assistant Secretary in 1685 and Secretary in 1713, Comptroller of the Chester Mint for 2 years from 1696, Savilian Professor of Astronomy at Oxford in 1703, and the second Astronomer Royal at Greenwich Observatory in 1720. Needless to say, his scientific interests extended far beyond comets, and in some respects it is rather ironic that his name is now associated in most people's minds with this one relatively minor aspect of his studies. He was ahead of his time in suggesting that meteors were objects entering the atmosphere from outer space and not something akin to swamp gas or lightning, as most people then believed. He investigated the acceleration of the Moon's motion and produced a table of lunar positions that led to the determination of accurate longitudes at sea. In addition, he investigated the periodic irregularities in the motions of Jupiter and Saturn and the proper motions of stars in the sky (comparing the positions of several stars with those determined by the ancient Greeks and noting the small changes that had taken place over time). He also used transits of Venus to measure Earth's mean distance from the Sun, produced the first general world chart showing variations of the compass, associated auroral activity and Earth's magnetic field, discovered the great globular star clusters in Hercules and Centaurus, and held modern ideas about the nature of nebulae. Moreover, he undertook ocean voyages (he commanded the *Paramour* for 2 years from 1698 on an expedition to measure magnetic variation) and, had he extended this somewhat further, would likely have discovered the east coast of Australia three decades before James Cook was born! On the downside, he also introduced goats to Trinidad and was therefore indirectly responsible for future severe erosion as the descendants of his animals did what goats do best to the local vegetation!

All this and more was accomplished by Halley during his active life. Somehow, he also found time to apply the theory of his friend Isaac Newton to the motion of comets.

In 1682, Halley married Mary Tooke and settled at Islington. In September of that year a bright comet appeared, and Halley was among those who observed it and carefully noted its changing position from night to night with respect to the background stars. At the time, this comet seemed no more or less interesting than any other bright one. Indeed, it was the magnificent apparition 2 years earlier that most held the interest of Halley and Newton. Only in retrospect did the comet of 1682 assume the full importance due to it.

Newton had used his theory of universal gravitation to compute the orbit of the Great Comet of 1680, but it was Halley who first applied the method to several comets that had been tolerably well observed during the preceding several centuries. Involving "prodigiously long and troublesome calculations" his catalog of the orbits of 24 comets was eventually published in 1705, and it was through his comparison of several of the orbits in this list that he came to the conclusion that at least some comets appear again and again. As we have already seen, he actually thought that his list included three returning comets, but the most promising was the one he observed himself in 1682. The orbit of this comet looked suspiciously similar to the ones he calculated for the comets seen by Kepler in 1607 and by Apian in 1531. These similarities could be explained most economically if these different comets were really just different apparitions of a single one moving in an elliptical orbit with a period of around 76 years. As a further tantalizing hint of this, Halley noted that a bright comet had also appeared in 1456 (though he had not investigated the orbit of that one) and even earlier comets had been noted at similar intervals. On these grounds, he predicted that the comet would reappear around 1758.

Although Halley died in 1742, the prediction was remembered, and in the year of the predicted return, a more precise calculation was performed by Alexis Claude Clairaut. Clairaut began with the 1607 apparition of the comet and painstakingly worked out its motion right through 1759, taking into account the strong gravitational fields of the planets Jupiter and Saturn. In what can only be described as a Herculean task in the days before computers and calculators, he determined the period of the comet at that time to be 76.85 years, setting the next perihelion date around mid-April 1759. This improved orbit narrowed down the search area, and on Christmas night 1758, the comet was first sighted by German farmer and amateur astronomer Johann Georg Palitzch, using a Newtonian reflector of 6 in. aperture and 8 ft focal length. Halley's hypothesis of a returning comet had been proven correct, and this object has ever since been known as Halley's Comet. Actually, with the naming of this comet in honor of its discoverer (or, at least, in honor of the discoverer of its periodic nature), the custom of naming comets for discoverers was established and, with minor modifications, has been continued ever since.

Following the return of 1759 and the confirmation of Halley's periodic comet hypothesis, the search was on for earlier and earlier appearances. Some of the suggested identifications were pretty wild, and others that at first appeared credible were later found to have been erroneous, as improved orbital computations became available. Each return of the comet since 1759 brought a new batch of possible ancient identifications, but it was not until the work of D. Yeomans and T. Kiang heralding the most recent return of the comet that the orbital elements were extended backwards with sufficient accuracy to permit reliable identifications to before the time of Christ.

The earliest identifiable return of Halley's Comet, based upon the work of Yeomans and Kiang, occurred in the year 240 B.C. (incidentally, in many astronomical texts the traditional year of Christ's birth is given as year 1 and the preceding as year 0. Earlier, years were given minus dates; thus 240 B.C. becomes −239 and 44 B.C. becomes −43 etc. The B.C./A.D. designation will, however, be used here, as it is more familiar to most people).

Astronomers actually calculated the orbit back much further than 240 B.C. In fact, they computed the comet's orbit right back to the year 1404 B.C., when the computed comet would have made a very close approach to Earth. The effect of such a close approach would have altered the orbit slightly, making it unrealistic to extend its motion back further in time.

You should note that the close approach in 1404 B.C. was by the computed comet, not necessarily the real one. Because there are no known observations earlier than 240 B.C., there is no way for the computed orbit to be continuously corrected against the real one earlier than that year, and in all probability the real and computed orbits had gotten somewhat out of sync by 1404 B.C. It is very probable that the real comet did not make an exceptionally close pass of our planet that year. A difference in the perihelion date of several days makes a big difference in how close an approach will be, and the uncertainty in the computed perihelion dates is large enough to make a very close passage uncertain.

Yeomans and Kiang also computed the orbit forward in time from the most recent apparition in 1986 and found that the comet will come to perihelion next on July 28, 2061, and again on March 27, 2134. On May 7 of that year, the comet will pass just 0.09 AU from Earth, as remarked earlier. Just two returns in the future, the hypothetical path of the computed comet will be pretty much the same as the actual path of the real one and (excepting the very remote possibility that Halley's falls to pieces in 2061), *this* predicted approach will really happen. However, the slight alterations in the comet's orbit resulting from its encounter with Earth make predictions into the more remote future too inaccurate for Yeomans and Kiang to extend their orbit any further.

But at least we now have a reliable history of the comet in past ages and can compare its appearances over a longer time than for any other comet. With this in mind, we are now ready to look at this famous comet through the eyes of our forebears, but before embarking on this journey into the past, we need to say a few words as to how the dates and times of the comet's perihelion passage are recorded in the following pages. The times of a comet's perihelion passage, as well as those of recorded observations, are given in Universal Time and noted in decimals of a day. Universal Time (UT), as its name implies, is constant throughout the world, irrespective of local time zones. Zero hours UT is midnight at Greenwich (England). From eastern Australia, zero hours is actually 10 A.M. local time (or 11 A.M. when daylight saving is operating). The decimal notation is simply the hour of the day given as a decimal of 24. Thus, 12 h UT on December 22 is written as Dec. 22.50 (half way through the day). If an observation was made on December 22 at 10 h 50 min UT, in terms of this notation it would be 10 5/6 of 24 = 10.83 of 24 = 0.45. The observation would therefore be recorded as having been made on December 22.45 UT.

240 B.C.

Perihelion date = 240 B.C. May 25.12. Closest approach to Earth = 240 BC June 3, 0.45 AU.
The Chinese observed the comet as a "broom star" sometime between May 24 and June 23, first in the east, moving to the north, and then in the west. Except for the implication of a tail in the classification broom star, nothing is said about the comet's appearance. Taking into consideration the meager data in the records and the comet's path according to the Yeomans–Kiang orbit, it seems that the comet was first seen in late May and again from late in the first week of June, as it emerged into the western sky.

164 B.C.

Perihelion date = 164 B.C. November 12.57. Closest approach to Earth = 164 B.C. September 28, 0.11 AU.
This appearance of Halley's Comet was not recorded in China and remained obscure until F. R. Stephenson, K. K. C. Yau, and H. Hunger found a reference to it, in 1984, on Babylonian tablets in the British Museum. By determining the date at which planetary configurations, also mentioned on the tablets, had occurred, they were able to date the observations and even restrict the time of

perihelion passage to a date between November 9 and 26 of that year. The Yeomans–Kiang orbit indicates November 12, in good agreement.

The comet was seen in Taurus, near the Pleiades, but no physical description is given.

87 B.C.

Perihelion date = 87 B.C. *August 6.46. Closest approach to Earth* = 87 B.C. *July 27, 0.44 AU.*

This apparition of Halley was also noted by the Babylonians, who saw it "day beyond day" through the lunar month of July 14 to August 11, and again on August 24, when it was said to have had a tail 10 degrees long. The only other observation comes from China, where it was recorded as a "sparkling star in the eastern quarter" sometime between August 10 and September 8. Its characterization as a "sparkling star" probably indicates that the tail was not very prominent at that time. From the consideration of the comet's orbit, it seems likely that the Chinese observation was in the early part of the period, as the comet headed rapidly into twilight by late August. However, it would then have been visible in the west, not in the east, leading Kiang to suggest that the direction given in the record is a mistake and should read "in the western quarter." In any event, it is unlikely that the comet remained visible much later than August 24 and was probably hidden in twilight before the full tail development had taken place.

12 B.C.

Perihelion date = 12 B.C. *October 10.85. Closest approach to Earth* = 12 B.C. *September 9, 0.16 AU.*

By contrast with the above apparitions, the Chinese probably documented Halley more fully at this return than any previous comet.

It was first recorded as a sparkling star in Gemini on the morning of August 26 and, between that date and September 9, passed through Lynx, Leo Minor, Ursa Major, and Canes Venatici. Passing Earth, it raced through Coma Berenices, Bootes, Serpens Caput, and into Ophiuchus during the following 10 days. It finally went out of sight on October 25 when in Scorpius and very low in the evening twilight.

There is also a mention in Dio Cassius' *Roman History* of a "star called a comet" that was said to have "hung for several days over [Rome] and was finally dissolved into flashes resembling torches [meteors?]" This is believed to refer to the 12 B.C. apparition of Halley's Comet.

Little is recorded, however, about the physical appearance of the comet at this return.

66 A.D.

Perihelion date = 66 January 25.96. Closest approach to Earth = 66 March 20, 0.26 AU.

At this return, Halley's Comet was apparently noted by the Chinese as a "guest star" as early as January 26. If that early observation was indeed the comet, its classification as a guest star probably indicates that in bright twilight only the central condensation was visible and that this looked more like a bright point of light than a typical comet.

No further reference to the comet was recorded until February 20, when it was again classified as a guest star but described as clearly cometary, with a tail of about 12 degrees in length. It seems not to have been associated with the object seen in January, although both fit the path of Halley's Comet. The guest star noted on February 20 remained visible until April 11, but although its track across the sky is well described, nothing further is said as to its appearance.

The Jewish historian Flavius Josephus writes that one of the portents of the destruction of Jerusalem by the Romans in 70 A.D. was "a star resembling a sword, which stood over the city, and a comet, that continued a whole year." From its association with the destruction of Jerusalem, A. A. Barrett dates this as 69 A.D., but from a reading of Josephus, it seems more likely that it occurred just prior to the Jewish/Roman war, which culminated in the destruction of the city. This would place the event in 66 A.D. and probably refers to Halley's Comet, as is frequently assumed.

But did Josephus refer to one comet or two? And did one of these remain visible for a full year?

As Halley's Comet approached Earth following perihelion, it was moving relatively slowly and may have appeared to "hang" over the city with a dust tail resembling a sword. Maybe, as this tail faded, its appearance changed to a more familiar cometary one, but it certainly did not remain visible for a year! It should be noted, however, that Josephus was not above a little exaggeration to make a point, and this might be another instance of his brand of emphasis!

141

Perihelion date = 141 March 22.43. Closest approach to Earth = 141 April 21, 0.17 AU.

The Chinese first noted this return of Halley's Comet on March 27, describing it as "pale blue" and with a tail of some 10 degrees pointing toward the southwest. As it approached Earth it brightened to perhaps magnitude –1 or better during the close sweep past our planet between April 20 and 22. At that time, it must have been a very conspicuous object in the evening skies. Because Earth was close to the plane of the comet's orbit in late April, the dust tail and ion tail would have appeared together and been very intense.

Rather strangely, although the comet reached its greatest elongation from the Sun (80 degrees) on May 7, it seems to have last been observed the previous day. If its brightness behaved as at other returns – and there is no good reason to suppose that it did not – the comet should still have been an easy naked-eye object.

218

Perihelion date = 218 May 17.72. Closest approach to Earth = 218 May 30, 0.42 AU.

Halley's Comet was again observed by the Chinese, being first recorded as a sparkling star appearing sometime between April 14 and May 12. It moved through the northern constellations of Aries, Perseus, Auriga, Lynx, Cancer, and Leo from May until early June, when it went out of sight. Nothing is reported about its appearance, just that it was first seen in the morning and then after more than 20 days reappeared in the western evening sky.

A possible Roman mention of the comet is also found in Dio Cassius' *Roman History*, where it is regarded as an omen of the revolt of Emperor Opellius Macrinus (commonly dated by historians as June 218). Cassius writes that "a comet was seen for a considerable period; also another star, whose tail extended from the west to the east for several nights, caused us terrible alarm." This may not refer to two objects, but to the morning and evening appearance of the same comet. By this description, it seems that the evening appearance may have been truly spectacular and the tail very long. Rather mysteriously, however, Cassius refers to a non-existent "eclipse" of the Sun just before the revolt, which tends to cast a shadow of doubt over his account of a comet. Maybe the eclipse referred to some meteorological phenomenon that darkened the Sun (a "dark day," for instance). Or maybe he simply invented portents if nature failed to supply any!

295

Perihelion date = 295 April 20.40. Closest approach to Earth = 295 May 11, 0.32 AU.

This was a rather poorly observed return of the famous comet. It appears to have been first seen by Chinese astronomers on April 30 and tracked until some time in June. The comet was variously described as a sparkling star or a broom star, but typical of early Chinese accounts, nothing beyond this is said describing its physical appearance.

374

Perihelion date = 374 February 16.34. Closest approach to Earth = 374 April 1, 0.09 AU.

This marks the second closest recorded approach of Halley's Comet to our planet. It must have been a remarkable spectacle, although there is nothing in the available records to suggest this, probably because it was better placed for the southern hemisphere at the time of minimum distance.

Chinese astronomers first saw it as a sparkling star on March 3, when it was located in the morning sky. On April 1, it appeared in the south as a broom star and reached an elongation of 166 degrees from the Sun on the third of that month. This must have been an incredible sight, but (once again) physical description is lacking. The comet went out of sight sometime during the month of April.

Although there is nothing in the very matter-of-fact description to suggest it, Halley's Comet at this apparition almost certainly deserves a place among the greatest of the greats. However, if we did not know the orbit and were not aware of its close approach that year, we would not have guessed this from the historic records alone. This should be remembered in any final assessment of which comets were and which ones were not truly the greatest in history. If we almost missed Halley's of 374, chances are we have missed others as well!

Incidentally, the Yeomans–Kiang orbit disproves any association with the sparkling star recorded by the Chinese on November 19, 374. This comet – if indeed it *was* a comet – must have been a totally unrelated object.

451

Perihelion date = 451 June 28.25. Closest approach to Earth = 451 June 30, 0.49 AU.
Once again, Chinese astronomers were first to see the comet at this return, finding it in the morning skies of June 10 near the Pleiades star cluster in Taurus. During the next month, it traversed the constellations of Perseus, Auriga, Lynx, Leo Minor, Leo, and Virgo, finally becoming lost in twilight on August 16 when located just 26 degrees from the Sun. Although no physical descriptions are given in the Chinese records, the comet probably reached around zero magnitude at the end of June.

530

Perihelion date = 530 September 27.13. Closest approach to Earth = 530 September 2, 0.28 AU.
Astronomers in China first noted this return of Halley on the morning of August 28, in the region of Ursa Major and displaying a tail of about 9 degrees. It was described as a broom star, suggesting that the tail had already become quite obvious.

The comet reached a maximum elongation from the Sun on August 16 (70 degrees) and a maximum northerly declination of + 43 degrees at the end of that month, after which it went out of sight in the morning sky.

The Chinese next saw it on the evening of September 4, when it was in the northwest with a tail 1.5 degrees long. The next day marked minimum solar elongation (27 degrees), after which it again began to pull away from the Sun (in terms of apparent separation, not in actual distance), reaching a new maximum of 38 degrees on September 16. All this while, it was actually approaching the Sun in real terms, but receding from Earth. By September 26, the comet had become barely visible in twilight, and the final observation seems to have been on the following evening, just as the comet arrived at perihelion. It was then at 32 degrees elongation from the Sun.

This return was also recorded in Byzantine texts, which describe a comet appearing in the west as a "tremendous great star ... sending a white beam upwards" and being visible for about 20 days. Some people were said to have called the comet "Firebrand" ("Lampadias" or "torch-like"). One Byzantine record mentioned that the tail extended "to the zenith." As the head must have been quite low when the sky grew dark enough for the extended tail to be seen, this implies a length of 70–90 degrees. Tail lengths of this order were certainly not implied in the Chinese record, however, and it is probably wise to be a little skeptical here.

607

Perihelion date = 607 March 15.48. Closest approach to Earth = 607 April 19, 0.09 AU.

This is the third closest approach to Earth by Halley's Comet in recorded history, only marginally more distant than that of 374. Actually, in the year 600, the *orbit* of the comet essentially intersected that of Earth, making for a potentially very close approach in 607. Calculations by Guy Ottewell show that if the comet had arrived at perihelion on March 10 that year, it would have approached to about two thirds of the Moon's distance from Earth! As it was, the comet passed just inside Earth's orbit near the time of closest approach, implying that we must have come very close to encountering its tail.

Unfortunately, the only records we have of this apparition (Chinese) are rather garbled. Although some Chinese records note that a "long-tailed star" became visible on April 4, others speak of a similarly long-tailed star that extended across the sky being seen as early as February 28 and of a broom star appearing on March 13 and lasting for more than 100 days. And, as if to throw more confusion into the melting pot, a sparkling star was recorded for June 25 as well as another comet beginning on October 21 and almost circling the whole sky during a 3-month period of visibility!

Various attempts have been made to sort all of this into some order, but it is no easy task. It is possible that some of the dates are seriously wrong, but that is probably not the entire story, as some of the positions given for the comet(s) and orientation of tails are not consistent with Halley at *any* date that year. The February/early March object does not appear to have been Halley, and unless the June observation is badly misdated, it must have been something else as well. The October comet cannot fit Halley either. Even if the dates are ignored, the positions and duration of visibility would seem to preclude it.

On the other hand, Stephenson and Yau suggest that the February date may have been a mistake, and that two relatively minor alterations in the Chinese text yield a date of March 20 as the first sighting. This is far more in accord with Halley's Comet. T. Kiang further suggests that a similar correction would turn the March 13 date into April 18, again bringing the record into better conformity with Halley.

Nevertheless, a problem with the April observations remains. The April 4 account places the long-tailed star in the west at a time when Halley should have been conspicuous in the *east*! Gary Kronk suggests that since the comet should certainly have been noticed (and presumably recorded) in the east at that time, this apparent discrepancy can be nothing more than a simple error of writing "east" instead of "west." If there really had been one great comet in the east and another in the west, we should indeed think that somewhere this would have been mentioned!

And yet, Stephenson and Yau worry whether the tail of the comet would have appeared long enough on that date (with the comet still 0.6 AU from Earth) to warrant the description of long-tailed star. We might also note that the "February 28" (read March 20?) observation also categorized the comet in the same way and went on to say that the tail "extended across the sky." We wonder, though, just how long is long in this description, and we may ask (but not expect an answer) just what "extended across the sky" *really* means. Right across? (Not very likely) Fifty degrees across? Ten degrees across? One degree across? It is a pity that such confusion exists in the records of what must have been one of Halley's finest performances.

684

Perihelion date = 684 October 2.77. Closest approach to Earth = 684 September 6, 0.26 AU.

This apparition of Halley is best known for its depiction in the *Nuremberg Chronicles* of 1493, where it was blamed for bringing 3 months of heavy rain and lightning, resulting in the deaths of numerous people and animals and the ruination of fields of grain.

From Chinese annals, we find that the comet was first observed on September 6 and again the following evening as a broom star located in the northwest with a tail of about 15 degrees. Although there is some confusion in the records, it is likely that the comet remained visible until October 8 as it sank ever lower into the twilight.

The comet is probably identical with the "hairy star ... having a pillar-like shining" recorded in Ghevond Yeretz's *History of Armenia* and the "great comet [seen] every evening for 41 days," said by Michael the Syrian to have appeared around September 684. It is also interesting to note that Michael mentions that this great comet was preceded by another "large comet" that lasted for 11 days. According to the orbit, Halley's should have been easily visible before dawn from the middle of August until the first days of September. Could this morning appearance have been the large comet to which Michael was referring?

Also of interest is Michael's further note that "others appeared alongside [the great comet] for seven days in the month of September." What were these "other" comets? Fragments from the nucleus would have remained too close to the central condensation to have been visible by eye, but it is possible that some rather extreme disconnection events in the plasma tail might have given the impression of secondary comets, although it is probably useless to speculate further with so little evidence available.

760

Perihelion date = 760 May 20.67. Closest approach to Earth = 760 June 2, 0.41 AU.
The comet was first noticed in China as a broom star with a 6 degree tail located in Aries before dawn on May 16. It then moved rapidly toward the northeast before passing into the evening sky. It was visible for a total of 50 days.

This appearance of Halley's Comet was also recorded by the Byzantine monk Theophanes the Confessor, who noted that the comet was "very bright" and appeared for 10 days in the east and 21 days in the west.

Arabic historian Agapius Manbij noted that in the year 760 a star with a tail appeared in Aries before sunrise, then "proceeded until it was under the rays of the Sun, then went behind it and stayed 40 days."

837

Perihelion date = 837 February 28.27. Closest approach to Earth = 837 April 10, 0.03 AU.
The return of 837 marks the closest known approach to Earth of this comet and among the closest for *any* comet in recorded history. It was also the most thoroughly documented account of Halley's Comet during the first millennium.

First recorded in China on the morning of March 22, it was then described as a broom star with a tail 10–11 degrees long. During the remainder of that month, the tail increased in both length and intensity, all the while pointing toward the west. On April 6, it was over 30 degrees long and possibly 5 degrees wide, and just 2 days later split into two branches and acquired a length of perhaps 75 degrees. By April 10, the two branches had come together, but the tail continued to lengthen to about 90 degrees, pointing toward the north. The next night, the tail was down to 75 degrees in length but about 7–8 degrees in width. The tail's length was possibly in excess of 100 degrees on the 14th, when it pointed toward the east, but had shrunk to just 4.5 degrees (still directed eastward) on April 28, when the comet was last detected.

The comet was also recorded in Japan, Germany, and in the Arab world and, although no extra information is added to the Chinese description, all records agree that it was very bright and sported an extensive tail.

As a point of interest, the comet swung far south at the time of its closest approach and actually followed a path through the constellations very similar to its track in 1986. There was one big difference, however. While its great southern excursion took weeks in 1986, it all happened within a couple of days in 837 as the comet sped past Earth at very close quarters. What an outstanding sight it must have been from the southern hemisphere, high in the night sky and just 0.03 AU away!

Imagine standing under a pristine sky somewhere in what was later to become Australia, southern Africa, or South America. Overhead is a great nebulous mass of light maybe 7 degrees across (that is to say, 14 times the breadth of the full Moon!) and with a total light between that of Jupiter and Venus, but most probably closer to the latter. Bright enough, that is, to cast weak diffuse shadows on a white surface. In the center of this mass of light, a star-like point shines at first magnitude or thereabouts, and a great tail extends almost to the horizon. At the time of closest approach, this tail is probably widened into a great fan of light, dominating the whole visible sky. As you watch, the head of the comet is seen slowly moving against the starry background. Although no such thought would have occurred to anyone in 837, the motion would be likened today to the minute hand of a clock. This was the spectacular, and probably terrifying, sight that must have been beheld by the aborigines of Australia, the Bushmen of southern Africa, and the various aboriginal peoples of southern America on that long-ago night of April 10, 837! Few of the greatest comets of history could have surpassed Halley's on that occasion.

912

Perihelion date = 912 July 18.67. Closest approach to Earth = 912 July 15, 0.49 AU.

Contrasting with the above apparition, the return of 912 was rather lackluster, although accounts of the comet are found in Oriental, European, and Arabic chronicles.

Japanese sources first record the comet as a broom star on July 19 in the northwestern evening sky. It remained visible only until July 28, and no further physical description is given.

On the other hand, Chinese annals record a broom star that appears to follow the path of Halley's Comet on the nights of May 13 and 15. With a perihelion date of July 18 well established, Halley's could not have been observed on these dates, and at one time it was even suggested that what the Chinese saw may actually have been a fragment that had broken away from the main comet and arrived at perihelion 2 months ahead of schedule! This raises the difficulty, however, as to why this hypothetical fragment was observed and not the main comet itself. Errors in dating are more likely, and I. Hasegawa suggests that the intercalary month was left out, implying that the dates should really have been July 12 and July 14, in conformity with both the Japanese observations and the comet's perihelion date of July 18.

989

Perihelion date = 989 September 5.69. Closest approach to Earth = 989 August 20, 0.39 AU.
The first people to see the comet at this return appear to have been the Swiss, who noted it in the western sky on August 10.

The Chinese noted it as a guest star on the 11th and as a broom star on the 13th and Japanese chronicles record a broom star with a tail of 7–8 degrees in mid-August. Although possible and probable references to the comet are found in the chronicles of several countries, little further information about its appearance is forthcoming.

1066

Perihelion date = 1066 March 20.93. Closest approach to Earth = 1066 April 23, 0.10 AU.
This is undoubtedly the most famous return of Halley's Comet prior to modern times. The story of its appearing before William the Conqueror invaded Britain and how it became incorporated into the Beaux Tapestry is well known. It was extremely widely observed, with mention being made in the annals of Oriental, European, Byzantine, and Arab cultures. There is even evidence that it found its way into the oral traditions of Hawaii!

The most extensive series of observations, however, were made by the Chinese. They were the first to notice the comet on the morning of April 2, and they kept track of it until June 7 – a remarkably long duration. Indeed, the computed orbit indicates that on the date of the last observation, the comet would have been 1.61 AU distant from the Sun and 1.69 from Earth. Under normal circumstances, it should have been, at best, a very marginal naked-eye object then, and is most unlikely to have been recorded by the astronomers of the day. It seems pretty clear that it was anomalously bright at the time, but it is not obvious whether this anomalous luster was present throughout the apparition (as Vsekhsvyatskij thought) or whether the comet experienced a large brightness outburst, similar to the one that occurred late in the 1835 apparition, as most comet experts now tend to think.

According to Oriental records, the comet was first visible before sunrise from early April and possessed a tail of around 10–11 degrees. It then passed into twilight and reappeared in the evening as a large and brilliant object from about April 24, at first without an obvious tail (presumably because of bright twilight) but soon developing a broad one over 20 degrees long. On the 25th, the tail was said to have been 15 degrees long and 4–5 degrees wide. The coma was obviously very large at the time the comet appeared in the evening sky, as descriptions from Korea, Italy, Byzantium, and the Arab world all compare it to the Moon. An Italian record says that it appeared like an eclipsed Moon with a tail that reached "half way to the zenith." Some descriptions suggest that it had double or even multiple tails. One Byzantine record mentions that it first appeared with a "smoky, cloudy tail" but later developed a "curly tail" as well. The smoky tail was probably a Type II dust tail, and the curly tail, a turbulent Type I plasma tail.

Whether or not the comet was anomalously bright at that time, forward scattering geometry should have enhanced its apparent magnitude around the time of minimum approach to Earth. An Egyptian record mentions that the comet "appeared during the last part of the day ... where the Sun would set during that month." Although not definitive, this suggests that it was seen in daylight at that time. Given the comet's proximity to Earth and a geometry favoring forward scattering of light by dust in the coma, this is not impossible.

In any case, the eleventh century return of this comet must surely rate as one of its brightest and most spectacular, as well as one of the more widely observed prior to modern times.

1145

Perihelion date = 1145 April 18.56. Closest approach to Earth = 1145 May 12, 0.27 AU.
Once again, this return of Halley's Comet was recorded in many lands. Once again also, it remained visible for an abnormally long time and either experienced another major outburst toward the latter part of its apparition or was more than normally bright throughout its appearance. It is not possible

to choose between these two alternatives with confidence. According to A. Pingre, the comet was first seen in Europe on April 15, when it would have been 0.58 AU from the Sun (just prior to perihelion passage) and 1.19 from Earth and, if behaving normally, shone at about second magnitude. At 29 degrees from the Sun, it should have been discoverable, but not very conspicuous. If it was already anomalously bright, it would have been an easier target, of course, but it should be noted that it was not seen in Japan before April 23, and not from China until April 25; this is quite consistent with something brightening from second magnitude. But even if the first observations cannot be used to distinguish between normal and anomalous brightness, the final ones clearly imply an abnormal luster. Japanese astronomers followed the comet until June 18 but were unable to locate it on the next clear night, June 27. The Chinese, however, last saw it as late as July 6, when it was 1.61 AU from the Sun and 2.1 AU from Earth. This was an even greater feat of observation than their predecessors had managed in 1066!

From discovery until early May, the comet was visible in the east before dawn as a broom star, white in color and with some 15 degrees of tail visible. It reappeared in the western evening sky around the middle of May with a tail of some 30 degrees, slowly fading and shrinking in size until finally going out of sight in early July.

1222

Perihelion date = 1222 September 28.82. Closest approach to Earth = 1222 September 5, 0.31 AU.
For this return of the famous comet, the honor of being the first to record it goes to the Koreans, who saw it on the morning of September 2, noting a westward-pointing tail of about 4–5 degrees. By the 8th this tail had grown to around 30 degrees.

The comet was first recorded in Japan on the 7th, and the following day its tail was said to have intensified and grown longer. Another Japanese record stated that on the 8th the broom star possessed a center "as large as the half Moon." Presumably this referred to the diameter of the entire coma rather than the central condensation as such.

There is no record of the comet having been seen in China before September 10, after which it was followed until possibly October 8, although there is some confusion as to the final Chinese observation. One record gives a date as late as October 23, while another implies a final observation around November 25. These dates are, to say the least, unlikely, as the comet was a mere 7 degrees from the Sun on October 23 and located beyond it!

Although mistakes in dating are the probable cause of the confusion about final observations, the biggest problem raised by the 1222 return is a Korean record for September 9 stating that the comet was seen "during the day" and that "Venus was seen during the day across the sky." That was the date of closest approach to Earth, but even granting this, and even granting some forward scattering enhancement, the comet should not have been bright enough to be seen in daylight!

The simplest solution would be to assume a mistake, or to change "during the day" into "during the twilight" or something of that nature. But might there not be something more interesting here than a simple mistake? And if so, what could it be?

We suggested that the comet may have been brighter than normal during its previous two returns. At the very least, it apparently experienced some large outbursts then. Is it possible that it was unusually bright in 1222 or (more likely) that it experienced another great outburst fortuitously at the time of closest approach to Earth? Admittedly, all other major outbursts recorded for this comet occurred *after* perihelion, but that does not mean that there is a law against one occurring while the comet is still approaching the Sun.

There are two statements in Japanese records suggesting that something may have happened around that time, although they by no means prove it. For a start, the Japanese record for the night of the 7th states that, although the comet itself looked white, the "rays" (tail) appeared red. This is strange, as prior to its perihelion, the tail of Halley is predominately a plasma tail and, where color is mentioned at all, it is normally either white or bluish. Red reads more like a dust tail. Normally, such a tail would not be very apparent this early, but is it possible that a dust tail may have developed prematurely following a strong outburst? On the other hand, maybe there was simply a band of haze in the atmosphere that night or the observer's eyes were bloodshot!

The other possible hint that something may have been happening is the statement the following night that "its rays and brightness became intensified." True, the comet was approaching Earth, but that would hardly cause a very great change over the course of 24 h, and we may doubt that a normal brightening of this type should be deserving of special mention. Of course, this is just speculation, but the very fact that an increase in luster was specifically recorded might imply that there was a *major* brightening, even one sufficient to raise the comet to daylight visibility!

1301

Perihelion date = 1301 October 25.58. Closest approach to Earth = 1301 September 23, 0.18 AU.

On this return, Halley's Comet was picked up in the eastern sky around the middle of September and observed most extensively by Japanese and Chinese astronomers, although records also appear in European chronicles.

The Japanese first saw it on September 15 and described it as a broom star with a tail of around 4.5 degrees. The following day, it was observed in

China and described as being "like Procyon" (in brightness? color?) with a tail of 7.5 degrees which "formed a straight line in the northwest direction."

By September 23, the tail had grown to over 15 degrees in length according to the Chinese, who then followed the comet through the northern skies until the evening of October 31.

1378

Perihelion date = 1378 November 10.69. Closest approach to Earth = 1378 October 3, 0.12 AU.

This is the return of Halley's Comet that Fred Schaaf calls the "pole vault," the once in 3,500-year apparition where the path of the comet passed very nearly over the north celestial pole. On the night of October 2 (the day before closest approach to Earth) the comet came to within 7.5 degrees of the pole. Probably shining at first magnitude and sporting a tail of 15 degrees or thereabouts at the time, for a brief while the northern hemisphere boasted a spectacular "Pole Comet" in addition to the familiar "Pole Star."

The Chinese first recorded the comet on the morning of September 26 and classified it as a guest star. By the end of the month, however, it was displaying a tail of 15 degrees and rapidly becoming circumpolar.

Cloudy weather in China apparently cut short the period during which this comet was observed by the official astronomers, and the final observation was as early as October 11. Nevertheless, although no further observations were reported from the Orient, records say that in Japan, prayers were being offered because of a comet from November 5 until November 15, indicating that it was probably still visible as late as mid-November.

1456

Perihelion date = 1456 June 9.63. Closest approach to Earth = 1456 June 18, 0.45 AU.

Chinese astronomers were the first to sight the comet at this return, locating it on the morning of May 27 as a broom star with a tail 3 degrees long. This extended to around 15 degrees by June 6, but the comet was not recorded again in China until June 22, when it had moved into the evening sky and sported a tail of 13–14 degrees or more. The tail was estimated as 10.5 degrees on June 28, but only 1.5 degrees on July 6, when the last observation was made from China.

The comet was well observed in Europe, where an important series of positional measurements was made by Paolo Toscanelli in Italy. He first saw the comet on June 8 and followed it practically every night until July 8, when he secured the final observation of the 1456 apparition.

Another European observer of this return was Georg Peurbach of Austria, who first observed it on June 10 (although he apparently knew of its existence as early as June 3) and measured the tail as around 10 degrees. He also attempted to measure the comet's parallax, the first time that such a thing had been tried. From the lack of discernible parallax, he estimated that the comet was "more than a thousand German miles" away.

An anonymous treatise, *De cometa*, written in 1468, recalled a comet appearing on June 6, 1456, as "very clear and brilliant" and possessing a tail 22 degrees long.

Records of the comet are found in Japanese and Korean chronicles, the former having first spotted it on May 31 and the latter on June 6. According to the Japanese, the comet's tail was 4.5 degrees long on the date of discovery.

The comet was also seen in the Arab world, but aside from its description as "a star with a very long tail," little information is given.

1531

Perihelion date = 1531 August 26.24. Closest approach to Earth = 1531 August 14, 0.44 AU.

For a long time, it was thought that this was the comet featured in woodcut illustrations in Peter Apian's book *Practica auff dz. 1532 Jar*, showing the tail pointing away from the Sun. Although Apian did observe this comet, it is now thought that the object shown in these woodcuts was actually the Great Comet of 1532, which he observed extensively in October and November of that year.

In common with Halley's previous return, the Chinese seem to have been the first to see the comet, locating it on the morning of August 5 as a broom star with a tail longer than 1.5 degrees. By August 16, the tail had grown to over 10.5 degrees. The comet was followed in China for a total of 34 days, implying a final sighting on September 8.

Japanese and Korean astronomers saw the comet in August, recording tail lengths of 7.5 and 15 degrees, respectively.

Likewise, German astronomer Johannes Schoner observed it that same month, noting that it first appeared "brilliant red-gold" but later waned "and exhibited a more and more ragged coma." The comet was apparently seen in Mexico, the first comet to be clearly discernible in several Mexican texts, but no additional information can be gleaned from that source.

1607

Perihelion date = 1607 October 27.54. Closest approach to Earth = 1607 October 29, 0.25 AU.

At this return, Halley's Comet was widely seen in China, Europe, and Mexico, with the Chinese sighting, typically, spanning the longest period.

They first noticed it on the morning of September 21 as a broom star, described as being "pale in color" and having a tail of 3 degrees length. No further descriptions are given, however, until the final observation of October 26, when the comet is again noted as a broom star located in the east.

It appears that the Mexicans may have tied with the Chinese for the initial observation, as their records state that a comet was first seen on September 21 and remained visible for several nights.

On September 26, Johannes Kepler was in Prague watching a fireworks display and, after the fireworks had ended, noticed a comet in the constellation of Ursa Major. He followed it, on and off, until October 26, carefully noting its position but recording little about its physical appearance.

1682

Perihelion date = 1682 September 15.28. Closest approach to Earth = 1682 August 31, 0.42 AU.

This time around, the first sighting of the comet was made from the New World. On August 24, Arthur Storer of Maryland in the United States discovered the comet but remarked that he could not see a tail "either by eye or else by prospective glass." The following day in Poland, however, Johannes Hevelius found it and noted a tail 12 degrees long. That same day the Chinese also located it, describing it as a broom star with a tail of about 3 degrees. It seemed that the length of tail observed depended critically upon local conditions at the time. The tail was recorded as 9 degrees by the Chinese on August 28 and 5 degrees when observed by John Flamsteed at Greenwich, England, 2 days later.

Flamsteed again saw the comet, this time displaying 10 degrees of tail, on September 2, and Edmond Halley noted it himself on the 5th as an evening object in Coma Berenices. On the 8th it was independently discovered in South Africa by Simon van der Stel. By the 19th that observer recorded that it had become "much brighter than before," even though northern observers were reporting a fading of the tail from mid-month. Already on September 10, Flamsteed could only detect 2 degrees of tail, and by the 12th recorded that it had become virtually invisible to the naked eye.

The final northern hemisphere observation was by Storer on September 22, and the final southern hemisphere sighting by van der Stel on the 24th of September.

An interesting observation of a "bright sector" or jet emanating from the nucleus was recorded by Hevelius on September 8. This feature was very similar to coma structures seen during the 1835 and 1986 returns of this comet.

1759

Perihelion date = 1759 March 13.06. Closest approach to Earth = 1759 April 26, 0.12 AU.

The evening of December 25, 1758, was a milestone in the history of cometary astronomy, being the first deliberate recovery of a periodic comet. The honor of being the first to locate a returning comet goes not to a professional but to a German farmer and amateur astronomer, J. G. Palitzsch. After confirming movement the following night, Palitzsch reported his find to C. G. Hoffmann, who officially confirmed the recovery on December 28.

Unaware of what had happened in Germany, Charles Messier made an independent recovery of the comet on January 21, 1759, at the Marine Observatory in Paris, and observed it until it was lost in twilight on February 14.

The comet emerged from twilight in late March and was picked up in the morning sky by de la Nux, on what is now known as Reunion Island in the Indian Ocean, on March 26 and by Gaston-Laurent Coeurdoux in India the following day. A tail of several degrees was noted by both observers.

As April began, the comet again became accessible to northern observers, with Messier seeing it at very low altitude on April 1 and noting that not only had it brightened since his previous observation in February, but that it had sprouted at least 25 degrees of tail. It was widely observed through the first half of the month from both the Orient and Europe, with brightness estimates of around first magnitude and reported tail lengths ranging from just a couple of degrees to 12–13 degrees or thereabouts.

During the latter half of the month, the comet's apparent motion increased as it approached Earth and sped southward. By April 25, it was moving at 16.6 degrees per day and sweeping through the constellations around the south celestial pole. Just after reaching its closest point to Earth, the comet passed through the Southern Cross and sailed virtually overhead as evening progressed for observers in mid-southern latitudes. So close to Earth, it must have been a beautiful sight, and it is rather strange that more descriptions do not exist. For instance, the comet did not even rate a mention in South African records!

During May, the northern hemisphere was afforded its best views of the comet, and several estimates early that month gave a brightness of at least first magnitude. According to de la Nux, the tail extended as far as 47 degrees on May 5.

The comet was last seen with the naked eye on May 26 by the Chinese and the final telescopic observation was on June 22, when J. Chevalier (Lisbon, Portugal) spotted it just 15 degrees above his local horizon. On that date, it had receded to 1.94 AU from the Sun and 2.06 AU from Earth.

1835

Perihelion date = 1835 November 16.44. Closest approach to Earth = 1835 October 12, 0.19 AU.

Gary Kronk points out that for this, the second predicted return of the famous comet, more predictions were made than for any other comet before or since, with only the 1986 apparition of Halley posing a serious rival!

The comet was recovered on August 5 by E. Dumouchel at Rome when still 1.96 AU from the Sun and 2.46 AU from Earth. Brightening slowly, it became visible to the naked eye September 19, according to H. J. Anderson in New York. By the end of that month, it had become clearly visible by eye alone and was starting to show signs of a tail.

On October 2, F. W. Bessel observed an emanation from the nucleus that was probably similar to the bright sector seen by Hevelius in 1682. This feature was observed by a number of people throughout the month and often likened to a flame attached to the nucleus.

By October 4, the comet's brightness was being estimated as first magnitude and a short tail had become visible. This was clearly seen with the naked eye by October 10 and increased in length to between 6 and 12 degrees by mid-month as seen by the casual observer. Nevertheless, experienced astronomers who knew the trick of averted vision (the practice of looking just to the side of the object being viewed) managed to trace it to far greater distances. Tail lengths of 20 degrees or more were noted using averted vision at that time. On the 14th, according to Bessel, the tail extended for nearly 28 degrees and one estimate as great as 45 degrees was recorded around October 17. It is interesting to compare these maximum lengths with the relatively modest estimates made by the earlier generation of astronomers who saw the comet under very similar circumstances during its 1378 apparition. It would seem that early Oriental astronomers were not as adept at observing very faint tail extensions and would probably not have recorded the extreme tail lengths noted for some of the great comets of more recent years. As a corollary to this, it probably means that those ancient Chinese records of comets sporting 50 or more degrees of tail imply that these appendages were relatively bright over much of their lengths and that the comets concerned truly rated among the greatest of the greats (as is our assumption in Chapter 3).

Meanwhile, back to Halley's in 1835...

In common with other late-year returns such as those of 1378 and 1682, the tail became less conspicuous as the comet moved from the dark skies of high northerly declination and headed into twilight.

With the New Year, some interesting changes were seen taking place within the comet's head. On January 16 and 19, W. R. Dawes (England) remarked that the comet appeared "exceedingly faint" and on the 22nd P. H. L. von Boguslawski (Poland) estimated it at magnitude 6. Yet, according to T. Maclear (South Africa), it appeared like a second magnitude star on January 25! Several observers noted a "halo" surrounding the central coma and extending as much as 15 min of arc. A faint tail about 41 min of arc in length was noted by Maclear on February 13, and naked-eye sightings of the comet were still being made as late as March 24. These bright magnitudes so late in the apparition, plus the general appearance of the comet at the time, indicate that a strong outburst occurred in late January and have been seen as circumstantial evidence that similar events may have taken place during the returns of 1066 and 1145 where the comet was also visible with the naked eye unusually far from the Sun.

The last observation of the comet during the 1835 return was made by Boguslawski in Poland on May 19, very low over the western horizon just after evening twilight. It was then 3.02 AU from the Sun and 2.73 AU from Earth, the most distant that this comet had been observed until then.

1910

Perihelion date = 1910 April 20.18. Closest approach to Earth = 1910 May 20, 0.15 AU.

Since the previous return of Halley's Comet, the astronomical world had acquired a powerful new observational tool – photography. The 1910 return of this famous comet would mark its first photographic recovery as well as the first time that it left multiple images for future generations to peruse.

The first recognized photographic observation of the comet was by Max Wolf in Heidelberg, who used a 72-cm (28-in.) reflector at 1 h photographic exposure to recover the 16th magnitude comet on September 11, 1909. Subsequently, however, faint images were found on plates taken at Helwan in Egypt on August 24 and at Greenwich in England on September 9.

Following recovery, the comet slowly brightened. The first visual observation was made by Professor S. W. Burnham who, using the 40-in. refractor at Yerkes Observatory, saw it on September 15 at around magnitude 15.5. It brightened to 12th magnitude by the middle of November, near 10 in mid-December and 9 by late January 1910. Most telescopic estimates gave the faint 7's or 8 during February; however, on the 11th of that month, Wolf managed to glimpse it with the naked eye under very favorable circumstances. By comparing the comet's intrinsic brightness at a similar place in its orbit in 1985, Dutch amateur astronomer Peter Bus estimated that the comet's total brightness was then around 6.5. Wolf's, however, appears to have been the only naked-eye sighting of the comet prior to its becoming overpowered by evening twilight around March 12.

After passing on the far side of the Sun, the comet eventually emerged into the morning sky, where it was first spied on April 8 by H. A. Howe in Denver, Colorado, by F. Sy (Algeria) on April 11, and on the 12th by E. E. Barnard at Yerkes. The latter observer estimated the brightness of the false nucleus as about magnitude 8.3. Although the total light of the coma was much greater, nothing was seen with the naked eye. In fact, Barnard did not see the comet with the naked eye until April 29, although it had already been spotted sans optical aid on April 16 by D. Costa (Rio de Janeiro) and on April 22 by John Tebbutt of Windsor, New South Wales, who described it as "not a conspicuous object" that, nevertheless, sported a tail of "two or three degrees." The comet's low altitude in morning twilight was the reason that people were having difficulty observing it, as its total brightness had apparently risen to around 2–2.5 by then.

A remarkable observation was made by Michel Giacobini at Paris on April 18 when, using a 10-cm (4-in.) finder telescope, he followed the comet into daylight for about 1 h after sunrise. Likewise, on April 22, J. Comas Sola repeated this feat by following the comet until after sunrise through the 38-cm (15-in.) Mailhat refractor at the Barcelona Observatory, and Howe continued to observe it for 15 min following sunrise on April 25 through a 51-cm (32-in.) refractor. At first sight, this may appear to give the lie to estimates of 2–2.5 magnitude for the brightness of the comet, but in fact it says more about the comet's *appearance* than its brightness per se. Halley's is notable for showing an intense and almost stellar central condensation, and it was due to the intensity of the latter rather than any exceptional brightness of the coma that allowed these experienced astronomers to see it after sunrise. During its next return, in March 1986, Rob McNaught and Gordon Garradd each followed the comet telescopically to within a minute of sunrise, even though it was then little brighter than third magnitude. Moreover, both observers felt that under slightly better circumstances, they might have kept it in view into daylight. With slightly greater brightness in April 1910, the Giacobini, Comas Sola, and Howe observations are understandable, if no less remarkable.

The comet became more conspicuous during May as it pulled away from the morning twilight early in the month (reaching maximum elongation from the Sun – 41 degrees – on May 7) and brightened as it neared Earth. The tail also lengthened dramatically. According to Barnard, it went from 17–18 degrees on May 3 to 53 degrees on May 14, 107 degrees on the 17th, and at least 120 the following day, dropping back to 29 on the 24th (when the comet had become visible in the evening sky, but under less favorable conditions), before extending again – under better conditions – to 54 degrees the following evening and 65 degrees on May 26. After that date, the tail contracted to 53 degrees on May 27, 47 on the 30th, 25 on June 1, and 15 on June 9. At this last date, Barnard described the comet as being "only faintly visible with the naked eye." His final naked-eye sighting was on June 11 between scudding clouds. No tail was then discernible.

Several days around May 19 were remarkable. At 5 h Universal Time on that date, the comet actually transited the face of the Sun. Although nothing was visible (not surprising considering the small size of the nucleus and the tenuous nature of the coma), the observational geometry at the time was about as good as it gets with respect to forward scattering of sunlight, and the comet must have experienced a tremendous increase in apparent brightness just before passing onto the solar disc and again upon leaving it. Of course, all this happened at next-to-zero elongation, but it does seem that the forward scattering enhancement was enough to increase the comet's brightness to daylight visibility close to that time.

As early as May 14, some enhancement was undoubtedly present, and on that date A. Borrelly, using the 15.9-cm (6.25-in.) comet seeker at Marseilles Observatory wrote that "It was broad daylight and the comet was still visible."

It was afternoon in eastern Australia when the transit happened, and in Hobart (Tasmania), J. B. Bullock spied the comet in full daylight through a pair of binoculars, some 3.5 h prior to the event. Apparently brightening rapidly, just 1 h later he saw it with the naked eye and noted that it possessed a tail (jet?) pointing upward. Remarkably, he followed it for a further hour by eye and, through the binoculars, tracked it right into the rim of the Sun!

At the same time, according to a letter to *The Hobart Mercury* dated May 30, 1910, Mr. John Cotton wrote that "a group of ladies" noted a haziness about the Sun from about noon until 2.30 P.M. Although this was undoubtedly a local haziness having nothing to do with the comet, Cotton's letter continues that "On placing themselves so that the Sun was obscured by an intervening chimney, they saw a small white spot slowly approaching the Sun. This disappeared as it reached the Sun's disc." This seems to have been about the same time as the Bullock observations and was probably another daylight sighting of the comet. The transit began at 1.28 P.M. local time in Tasmania.

On the other hand, a reported sighting through smoked glass by an observer in Egypt gives an inconsistent position, and a number of reports of rapidly moving bright objects near the Sun were undoubtedly nothing more exotic than drifting baby spiders and thistle down.

Recent studies of forward scattering in comets at very small forward-scatter angles by Dr. Joe Marcus indicate that brightness enhancement may be extreme. Earlier work cited by Richter in the 1960s also suggested that increases as great as 10 magnitudes are not impossible.

In 1986, Reinder Bouma and the present author estimated the false nucleus of Halley as about magnitude 8 at a time when the comet was about the same distance from the Sun as it was in mid-May 1910. Given the far smaller distance from Earth at the time of the 1910 transit, the false nucleus should than have been about magnitude 4, without any help from forward scattering. Now, assuming that the latter really did amount to a 10-magnitude increase, the false nucleus of the comet should have been around –6 at the time of the Tasmanian sightings! These estimates may actually be conservative as some of the innermost coma could have been involved as well.

Not only the brightness but also the tail length was prodigious, as we can see from Barnard's estimates on May 17 and 18. Even these might be conservative, and Barnard himself was convinced that in addition to the bright tail (which then appeared as a long, tapering cone), another very broad and fainter glow to the south constituted a second tail so long that it disappeared beneath the southern horizon. This one was also seen by Mary Proctor and drawn as almost as wide as the Great Square of Pegasus.

On the night of May 19–20, a tail was visible in both the morning and evening skies and M. Antoniadi noted that from the tip of the morning tail to the tip of the evening one could be interpreted as a single tail some 240 degrees long! He even suggested that under very favorable skies, a continuous band might be visible connecting the two, effectively giving a tail spanning the full 360 degrees of sky. Remarkably, something of this nature may really have been seen from a ship in the Mediterranean on the morning of May 19. A passenger claimed to have seen a glow like an oversized Gegenschein opposite the comet. This glow was described as being some 45 degrees high and 50–60 degrees wide and to have a "pillar of light" at its center. This phenomenon may have resulted from sunlight backscattered from a portion of the dust tail outside Earth's orbit and lying directly opposite the Sun.

Did Earth pass through the tail?

Calculations by Z. Sekanina suggest that the edge of the dust tail may have just missed us – passing by at a little less than the distance of the Moon. The plasma tail, on the other hand, very likely did envelop our planet for a short time, but does not appear to have given rise to any discernible effects.

Fig. 2.1 Halley's Comet and Venus May 1910 (courtesy, Lowell Observatory)

Photography captured many of the phenomena reported and drawn by observers at previous returns of this comet, including bright jets of material from the nucleus region and spectacular disruptions within the plasma tail. The great outburst of 1836 was not repeated, however, although several small variations in brightness were reported, and a small flare resulting in a significant brightening of the false nucleus was noted by Tebbutt on the evening of May 30. He remarked that "the contrast between its appearance then and on the preceding and following evenings was very remarkable."

The comet fell below naked-eye visibility in June, but remained visible in large telescopes until the following year, with the last definitive photographic observation being taken at Flagstaff, Arizona, on May 30, 1911, when the comet was 5.23 AU from the Sun and glowing feebly at about magnitude 18. A possible image on a plate taken with the 36-in. Crossley reflector on June 15 remains uncertain.

1986

Perihelion date = 1986 February 9.66. Closest approach to Earth = 1986 April 11, 0.42 AU.

The most recent return of Halley's Comet was an event like no other in the history of cometary astronomy. Not only was the comet subjected to unprecedented scrutiny by Earth-bound techniques, but in March of 1986 it was met by an international flotilla of spacecraft from Russia, Japan, and Europe. The European craft, Giotto, made an especially spectacular – not to say hazardous! – trip through the central coma and secured the first ever close-up images of the nucleus, the true solid heart of a comet. Although it seems almost incredible now, until the Giotto flyby of Halley on March 14, 1986, there were still astronomers who denied that such a thing as a solid cometary nucleus even existed!

The comet was recovered by D. C. Jewitt and Danielson on October 16, 1982, at Palomar Mountain when still 11.1 AU from the Sun, 10.9 AU from Earth, and glowing very feebly at magnitude 24.2. There was no coma, and the tiny speck of reflected sunlight really was the nucleus that Giotto later imaged close at hand. A faint coma was detected by CCD imaging on September 25, 1984, when the comet was at 6.1 AU from the Sun and 6.2 from Earth, and the first visual observation was made on January 23, 1985, by Steve O'Meara using the 61-cm (24-in.) telescope at Hawaii. At that time, the comet appeared as an essentially stellar point of light of magnitude 19.6 surrounded by a very faint haziness.

The first naked-eye sighting came on November 8,1985, when Charles Morris and Steve Edberg drove to a dark site in California and were rewarded for their trouble by a glimpse of a faint patch of misty light about as large as the full Moon. Except for a few weeks around the time of perihelion, when the comet was behind the Sun, it remained a naked-eye object until May 29–30, 1986, when Richard Keen and Charles Morris in the United States and Andrew Pearce in Western Australia last detected it without optical aid as a very marginal hazy spot.

This is an unprecedented period of naked-eye visibility for Halley's Comet. In part it was simply that a special effort was made to see it for as long as possible, but the nature of the 1986 apparition also made possible the unusual length of naked-eye visibility. The return was an early year one, with closest approach following perihelion and favoring southern latitudes. However, it appeared *so* early in the year that the return actually had some of the characteristics of a late-year apparition as well. Recall that late-year returns see the comet favoring northern latitudes and making a close approach to Earth prior to perihelion, and in 1985 the comet did indeed favor the north to a certain degree and did actually make a relatively close approach to Earth (0.62 AU) on November 27. This is the closest secondary approach ever recorded for this comet. It was thanks to this combination of a "modified late-year" return with an early year one that enabled Halley to be followed for so long during its bright phase. Not only did this enable a long naked-eye window, but it also allowed for a longer period of scrutiny by powerful instruments in a way that most apparitions of the comet would not. How fortunate to have such a return just when we were ready for it!

The downside, however, was for the primary close approach to Earth to be rather more distant than many other early year appearances and for it to take place relatively late in the apparition when the comet had retreated beyond Earth's orbit and become intrinsically less bright and active than at most other times of minimum approach. All told though, this was a minor inconvenience!

The plasma tail began to develop as early as July 12, 1985, although it was only apparent on photographs then. By December 9, a double tail with components 3 and 4 degrees long was observed, and a faint tail several degrees long could be observed without optical aid as the comet moved toward evening twilight in late December and January. The first major disconnection event of the plasma tail occurred from January 9 to 11; a foretaste of what was to come! Similar major events were recorded on March 9–11 and April 11–12 with minor disconnections occurring frequently in March and April.

Following perihelion, the comet emerged in the morning sky in mid-February and became a relatively bright naked-eye object before dawn during the latter part of February and throughout March. During this period, the comet's brightness appeared to fluctuate very slightly from night to night, yet remain within the 2.6–2.9 range, and tail lengths of up to 12 degrees were visible with the naked eye. In common with previous returns, the central regions of the coma were active, with jets and fans of light emitting from the false nucleus.

Photographs around February 22 revealed a real surprise – a series of tails, the majority of them dust, spreading out in a spidery formation lying between the main dust tail and a short spike of an anti-tail projecting to the side of the coma. At least seven tails were counted on the initial photographs. Depending upon what one would classify as a tail, some photographs of the comet in late February could be said to reveal a grand total of 13 tails, not even counting the proliferation of rays within the plasma tail. Spectacular photographs from Siding Spring Observatory in New South Wales revealed rays splitting into sub rays and multiplying in a fantastic array! On March 25, as observed through binoculars, the ion tail was seen to kink almost at right angles about 3 degrees from the head, dramatically parting company with the diffuse dust tail, which maintained its (more or less) anti-solar orientation.

Fig. 2.2 Halley's Comet on March 8, 1986, showing a spectacular multiple tail structure, as photographed with the 1-m Schmidt telescope at the European Southern Observatory in La Sila (courtesy, ESO, and NASA)

At closest approach on April 11, the comet appeared as a large globule, larger than the full Moon and with a bright star-like core. Total brightness was around second magnitude, and the tail appeared as a very broad and diffuse fan that was difficult to trace with the naked eye. Part of this difficulty was the orientation of the tail, right across the brightest regions of the Milky Way!

On the evening of April 14, the comet appeared in the eastern sky with three tails pointing downward. This author estimated the brightness of the coma as 2.3 on that night, noting that the tail appeared as a faint veil around 4 degrees long. One observer on that night likened the comet to a shawl hanging in the sky. This is not a bad description...or was it a ghostly celestial jellyfish trailing its long tendrils toward the eastern horizon?

Fig. 2.3 Halley's Comet and the Milky Way, March 21, 1986 (courtesy, Terry Lovejoy)

By April 16, the delicate tail had increased to around 10 degrees, and on following nights up to 20 degrees could be traced shortly after midnight when the comet rode high in the sky. The tail then appeared as a thin gossamer veil becoming very broad at the extremities, and the similarity to an old woodcut of the Great Comet of 1811 reproduced in an astronomy text was striking.

Moonlight drowned the tail later in the month, but fortuitously the night of full Moon (April 24) was also one of a total lunar eclipse. At that time, the comet was almost overhead and still as bright as magnitude 3.5.

As the sky darkened, the tail of the comet just seemed to be drawn out longer and longer. At this author's location, it could be traced for at least 15–25 degrees naked eye, but under darker skies, Terry Lovejoy in Queensland managed between 42 and 46 degrees (the first 15 visible at first glance), with a suspicion of a very faint extension out to possibly 60 degrees. Actually, several observers suspected lengths of 60 degrees, and there was at least one of 100 degrees!

Were these very great lengths real? Lovejoy's 42–46 estimate is reasonable, but anything longer than 50 degrees or so remains in doubt. Gordon Garradd drew attention to the tail's orientation very close to the zodiacal band and suggested that the extreme lengths may have been of the band rather than the

Fig. 2.4 Halley's Comet, March 20, 1986. This photograph is close to the naked-eye appearance of the comet at the time (courtesy, Terry Lovejoy)

tail itself. If one blended into the other, it may have been hard to tell where one stopped and the other started.

The comet continued to fade through May, and the tail grew increasingly difficult to see. The plasma tail was last observed, on a photograph, on June 14, but the dust tail, remarkably, persisted very faintly until April 1, 1987, when it was last recorded on a CCD image. The comet was last observed visually on February 23, 1988, when David Levy saw it through a 1.6-m (61-in.) telescope at an estimated magnitude of 17. Traces of a dust coma were recorded in May 1988 but had disappeared by February 1990 according to a CCD image taken by Richard West, showing the comet as a mere point of light at magnitude 24.4. It seemed that Halley had finally shut down until circa 2060.

Well, not quite!

In February and March of 1991, the comet experienced a powerful outburst – at around 14.3 AU from the Sun! For a time, its brightness surged some 6 magnitudes – about 250-fold – peaking at magnitude 19!

After this brief burst of activity, it appears to have settled down again and had faded to magnitude 26.5 when imaged on January 1, 1994. As far as we know, the most recent observation was on March 8, 2003, by Belgian astronomer Olivier

Hainaut. At that time, the comet shone at an incredibly faint magnitude 28 and required simultaneous imaging by three of the four 8.2-m (323-in.) reflectors of the Very Large Telescope on Cerro Paranal in Chile. The present plan is to take an image of the comet every 15 or 20 years through aphelion, until it starts to brighten once more and becomes increasingly accessible to a future generation of telescopes on the countdown to its 2061 perihelion passage.

    This, then, is a short history of Halley's Comet through the ages. As we can see, it seldom made the ranks of the greatest of greats, but because of its regular appearances at relatively short periods it is the one most widely known and, in a sense, the one against which other comets tend to be measured.

    Nevertheless, there are bigger cometary fish out there, and the time has now arrived for us to have a look at them!

Chapter 3
The Greatest Comets of Ancient Times

Ancient people must have been fascinated, and maybe even terrified, by the appearance of bright comets in the sky, and it is not surprising that references to these objects turn up from time to time in the most ancient of writings. However, these very early accounts are often vague and not infrequently mixed with legend and fable. As evidence that people of ancient times took note of comets, they are important, but as accounts of individual comets themselves, these old stories yield little useful data.

The first comet to have been reliably recorded in more than one historical text is the one observed by the Chinese in the year 613 B.C. It was described as a "broom star" and said to have entered the Pei-Tou (i.e., the Big Dipper or Ursa Major) sometime during the period of August 4 to September 2, as we would reckon months. Nothing is specifically said about its appearance, and there is no indication that it became especially brilliant.

In fact, the first comet that we can include with confidence in our list of the greatest of greats has never been found in Chinese records. It *may* have been recorded, of course, but if so the records were presumably lost during the turbulent times into which Chinese history moved during the following centuries. Be that as it may, it is to the Mediterranean that we will shortly go for details of our first entry. But first, a brief word about the naming of the objects of our story.

Some of the very ancient objects that we will describe lack official names and designations. We will refer to, e.g., "Aristotle's Comet," but this is not an official name on a level with "Halley's Comet" or the like. For those objects that were relatively well observed and for which a discovery date is known, the designation first introduced in 1995 and retrospectively applied to all past comets is used. This system gave letters and numbers to comets according to the half-month of the year and numerical order of their discovery. For example, the first comet discovered during the first half of January 2007 was designated 2007 A1, the second 2007 A2, and the third 2007 A3. These three were found on January 9, 10, and 11, respectively. The next discovery took place on January 17, in the second half of January, and was therefore designated 2007 B1. In addition to these designations, comets are also identified in this system by means of a prefix, C/ for a comet of long or undetermined period, P/ for one of short period

D. Seargent, *The Greatest Comets in History*,
DOI 10.1007/978-0-387-09513-4_3, © Springer Science+Business Media, LLC 2009

(less than 30 years), D/ for a "lost" comet (generally a short-period comet that has failed to be recovered and is either of very uncertain orbit or has become defunct or dormant), and X/ for an object of such uncertain observation that an orbit has not been calculated. When a short-period comet is recovered and its orbit thereby secured with a good degree of accuracy, it is given a number. Thus Halley's Comet, being the first periodic comet whose return was successfully predicted, is known officially as 1P/Halley, Encke's Comet as 2P/Encke, and so forth. Not all of these (e.g., Halley) have periods of less than 30 years. If a C comet is shown to be a return of one seen many years ago, it is re-designated as P and numbered accordingly, irrespective of its orbital period. A good example of this is C/2002 C1 Ikeya-Zhang, which was found to be a return of C/1661 C1 and is now officially known as 153P/Ikeya-Zhang.

Comets, in keeping with a long tradition, are also named for their discover(s), although today the "discoverer" is often a robotic sky patrol and a large number of comets bear the names of programs rather than individuals.

Looking again at the above-mentioned comets of January 2007, 2007 A1 was an accidental recovery of P/1986 W1 (Lovas), 2007 A2 was also found to be periodic and has become P/2007 A2 (Christensen), A3 was a deliberate recovery of periodic comet Petriew, seen previously only at its discovery return, and 2007 B1 also turned out to be of short period and is now designated P/2007 B1 (Christensen) (E. J. Christensen's second discovery of the month!). Because both A1 and A3 were being observed at their second returns, they have now been officially designated as 184P/Lovas and 185P/Petriew. The first comet of 2007 bright enough to be seen in small telescopes at the time of its discovery was found by amateur astronomer Terry Lovejoy on March 15 and designated C/2007 E2 (Lovejoy). The prefix "C" indicates that it is not a short-period comet.

With these preliminaries in order, let's take a look at some of these fascinating objects.

The Great Comet of (ca.) 372 B.C., Aristotle's Comet

The great Greek philosopher and pioneer scientist, Aristotle, mentions four comets in his book *Meteorologica*, written around the year 330 B.C. In his mind, one of these apparently stood out from the rest and was always referred to as the "Great Comet." Such a phrase probably carried more weight in this ancient work than it does today, when great comet has become a semi-technical term for a large comet. One of the other comets Aristotle mentions (the one recorded by Anexagoras in 467 B.C.) was almost certainly great in our sense of the term, but was not designated as such by Aristotle. Anexagoras called this an "object of extraordinary grandeur," and many years later, the Roman historian Lucius Annaeus Seneca described it (presumably from original sources now lost) as having been "the size of a great Beam." Although this object appeared before Aristotle's birth, he must have been aware of these descriptions, yet apparently did not put it in the same class as the one of 372 B.C.

Concerning the latter, Aristotle writes in Book 1 of his *Meteorologica*:

> The great comet, which appeared about the time of the earthquake in Achaea and the tidal wave, rose in the west.... The great comet ... appeared during winter in clear frosty weather in the west, in the archonship of Asteius: on the first night it was not visible as it set before the sun did, but it was visible on the second, being the least distance behind the sun that would allow it to be seen, and setting immediately. Its light stretched across a third of the sky in a great band, as it were, and so was called a path. It rose as high as Orion's belt, and there disappeared.

Some translations of this record give "great ribbon" or even "great leap" for "great band," but the imagery of the band or ribbon of light is graphic. And what exactly does "It rose as high as Orion's belt, and there disappeared" mean? What rose? The comet (over a period of time) or the tail?

The final sentence is ambiguous as it stands, but because the immediately prior subject was the comet's tail rather than the comet itself, it is presumably the length and direction of tail that Aristotle describes, not the track of the comet through the constellations.

Other reports of this comet have come down to us, albeit secondhand, from later writers quoting earlier sources that have since been lost to history. For example, writing in the first century before Christ, Diodorus Siculus writes that "during the 102nd Olympiad, when Alcisthenes was archon of Athens...there was seen in the heavens during the course of many nights a great blazing torch which was named from its shape a flaming beam. Some of the students of nature ascribed the origin of the torch to natural causes, voicing the opinion that such apparitions occur of necessity at appointed times, and that in these matters the Chaldeans in Babylon and the other astrologers succeed in making accurate prophecies." Diodorus then went on to make the astonishing claim that "this torch had such brilliancy ... and its light such strength that it cast shadows on the earth similar to those cast by the moon."

As we shall later see, there have been other occasions where shadows have been reportedly cast by comets, but nowhere else does there seem to be a reference to these being comparable to those cast by the Moon. A degree of exaggeration is quite probable, but even granting this, the intensity of the light from the comet must have been fantastic. We also note how the comet itself (the head) is not specifically mentioned in any of these records and seems to have been quite inconspicuous in comparison with the great tail. It is the "blazing beam" of a tail that seems to have so impressed the observers of those far-off days and which came down in records over centuries of time. The shadows cast were most likely by the tail, which must have maintained an incredible intensity over much of its span.

This assessment of the comet as being overwhelmed by its tail is supported by an account written by Seneca about the year 63 A.D. Seneca wrote that

"Callisthenes reports that ... an extended fire appeared just before the sea covered Buris and Helice. Aristotle says that this was not a Beam but a comet. Moreover, he says that because of its excessive brightness the fire did not appear scattered but as time went on and it blazed less it recovered the usual appearance of a comet." Seneca further remarked that a Beam (presumably an auroral beam) "has an even flame" and is not "collected in the end parts" as this one was or, at least, as it became upon fading.

The immediate point of interest here is the remark that it "recovered the usual appearance of a comet." What seems to be implied here is a growing prominence of the head as the intensity of the tail waned. We may picture the comet as it first appeared; a long and brilliant tail ending in nothing more than a bright point at the lower end. So intense may we suppose the tail to have been that the head appeared as little more than its terminus. Later, however, as the comet moved away and the tail faded, the head swelled into a more typical cometary coma and became clearly differentiated from the dimming tail.

If anyone reading these words was actively observing back in 1965, when Ikeya-Seki appeared, would this description seem at all familiar? It sounds very like Ikeya-Seki writ even more brilliant!!

That comet also sported a very intense tail, against which the head was initially little more than an inconspicuous dot of light. Later, however, as the tail lost much of its intensity, the head took on a more prominent aspect and the object "recovered the usual appearance of a comet."

This appearance of a great band with only a small head is typical of bright Kreutz sungrazers (such as Ikeya-Seki) after perihelion. This family of comets is so interesting that Chapter 6 has been reserved specially for it in the present book. We simply mention at this time that the 372 B.C. comet has been named as a possible candidate for membership of the family. It has even been suggested that it might have been the parent comet from which the entire family descended through progressive schisms throughout intervening centuries. If that is true, the sungrazers of modern times (including the myriads of minicomets found by the SOHO spacecraft) might literally be pieces of Aristotle's Comet!

Interestingly, although no reliable orbit can be calculated for this object from the meager records that have come down to us, eighteenth-century cometologist A. G. Pingre did manage to compute some very uncertain orbital elements and hinted at a "very small" perihelion distance for this comet. Incidentally, the sungrazing group had not been recognized in Pingre's day.

A further statement of Seneca's is also of interest in this context. In an attempt to discredit the credibility of Greek historian Ephorus, Seneca refers to a (now lost) statement by this historian that the comet "split up into two planets." Apparently, Seneca thought this so absurd that its mere suggestion would be enough to discredit anyone making the claim. We now know that comets do sometimes split apart, and, indeed, the existence of the Kreutz group is itself proof of this. If Aristotle's Comet was the Kreutz's parent, or even just one of the early siblings, it probably did split while under observation in 372 B.C. However, separation of split nuclei takes place so slowly, that it is very unlikely that anyone would have recognized it. If the comet really did split in the manner that Ephorus recorded, something other than a simple schism of the nucleus must have been witnessed. Without further details, however, there is little point in speculating as to what may (or may not) have been seen.

Before leaving this fascinating comet, it might be interesting to speculate on the possibility that it had a profound, if indirect, influence on the later development of human thought. At the time of its appearance, Aristotle was still just a boy. His father was a physician and (although this is pure speculation) it might have been expected that the son would follow in the same profession. Yet, for some reason, Aristotle became interested in the natural sciences, philosophy, and cosmology. Is it possible that this interest was aroused by the sighting of an incredibly spectacular comet at an age when his life interests may have been germinating?

If this is true, the profound influence exercised by Aristotelian philosophy on the thought of the Western and Islamic worlds may have been an unexpected legacy of the comet of 372 B.C.!

Comet 135 B.C. (X/-134 N1), "Mithridates Comet"

In the opinion of modern-day cometologist Gary Kronk, this comet may have been one of the most spectacular of ancient times. From the records available to us, this opinion seems well founded.

Between August 31 and September 29 in the year 135 B.C., the Chinese record the appearance of a "long-tailed star" appearing in the east. They note that it remained visible for 30 days and that its tail stretched "across the heavens." Although this does not necessarily mean that it reached from horizon to horizon, it certainly implies a very great length that, as we shall shortly see, was also strongly hinted in the records of other ancient civilizations.

A later Chinese text records a long-tailed star in the year 134 B.C. (almost certainly the same comet misdated), and there also exists a report of a "sparkling star" from July 3 to August 1, 135 B.C., which may refer to earlier observations of the comet before its tail had become readily visible. As an indication of the reaction to the appearance of this comet, we may note an entry in a Chinese chronicle for the following year to the effect that the reign-period changed because of a comet!

Rome was the other ancient civilization to record a great comet about that time. Despite the tendency of some to differentiate between the Roman and Chinese comets, the chance of two such brilliant objects with unusually long tails appearing so close together in time seems too remote to be accepted without very good supporting evidence. Unfortunately, the existing Roman records all date from a time much later than the comet itself, and clearly depend upon earlier accounts that have since been lost.

The most sober Roman record comes down to us from the historian Lucius Seneca. He writes that some time during the reign of Attalus III, king of

Pergamum, "a comet appeared, of moderate size at first. Then it rose up and spread out and went all the way to the equator, so that its vast extent equaled the region of the sky which is called the Milky Way." Attalus III reigned from 138 B.C. until 133 B.C., placing the comet close to the time of the Chinese object. The identity of the two is strongly supported by the great tail length noted in the two countries, as well as by the lack of any competing comet recorded in Chinese annals dating from that period.

It would have been helpful if Seneca (or the record he was quoting) had mentioned the position of the comet's head, as then a better idea could have been formed of the tail's true extent. Nevertheless, his comparison with the Milky Way is surely telling. As noted by the Chinese, the tail must have "spanned" a goodly portion of the entire dome of sky.

A rather less sedate account is supplied by Seneca's fellow citizen, Marcus Junianus Justinus. Writing in the third century, Justinus noted that at the birth of Mithridates VI Eupator, "a comet burned so brightly for 70 days that the entire sky seemed to be on fire. In its greatness it filled a quarter of the heavens, and with its brilliance it outshone the sun, while its rising and setting each took a period of four hours." The birth of Mithridates has been determined as sometime between 134 and 132 B.C.; however, the slight discrepancy in dates should not worry us too much. Justinus may not have been above adjusting the dates by a year or so to make a better story!

Lest you think this latter remark cynical, let us point out that Justinus gives *exactly* the same account of a comet supposedly seen in the year Mithridates became king of Pontus, viz. in 120 B.C.! There *was* a comet that year, recorded on a cuneiform Babylonian tablet now designated as BMA 41131. It was said to have appeared in May and is most probably the same as the sparkling star recorded in China in the spring of the same year. However, there is no indication that it was even remotely as spectacular as the earlier one. It seems that Justinus merely rehashed the account of Mithridates' birth comet to also coincide with the beginning of his reign. If he was aware of the real comet of that year, perhaps he thought it too insignificant to commemorate such an auspicious occasion!

In any event, it is clear that Justinus' account is not without exaggeration. Obviously the comet's supposed brightness is out of the question. No comet can be brighter than the Sun!

However, the tail occupying a quarter of the heavens implies a length of at least 45 degrees and the force of occupied may also indicate considerable width as well. He did not say that the tail spanned a quarter of the heavens, as if length alone was in his mind. Perhaps he meant that the tail was of such vast dimensions that it effectively took up a quarter of the *area* of the celestial dome!

The statement that the comet took 4 h to rise and set is also very interesting. In itself, the time taken to rise and set implies a long tail, in conformity with the other surviving accounts. But the mention of it rising *and* setting would seem to indicate that it had a high northerly declination and was visible both after sunset and before dawn. (It may be noted here that the Chinese sparkling star was said to have been in the north.) Alternatively, it might have first been seen as a morning object, rising ahead of the Sun, before passing through into the evening sky and setting after it (or, perhaps, vice versa). A high northerly declination is the most probable explanation and gives added force to Seneca's statement that the tail "went all the way to the equator."

It is also worth noting the period of visibility given by Justinus was 70 days. This actually fits quite well with the suggestion that the Chinese sparkling star of July was indeed this comet. If it became visible around July 21 (which is within the period given by the Chinese account) and remained visible until September 29, its duration would indeed have been 70 days.

Moreover, Seneca's statement that the comet was "of moderate size at first" implies that it must have started out relatively small and without much visible tail. At that time, it could easily have been recorded in China as a sparkling star and not associated with the grand comet that burst onto the celestial scene 4 or 5 weeks later.

It seems that Mithridates himself was aware that the year of his birth coincided with that of an unusual comet as, during his reign, he had some bronze coins struck depicting a type of comet known to the ancients as the *hippeus*, or "horse" comet. The Roman naturalist Pliny described these comets as having plumes like horses' manes, most probably referring to the dust striae described in Chapter 1. The "horses' mane" appearance may have been similar to the tail of Comet McNaught in 2007, and if the images on Mithridates' coins are accurate in their depiction of the comet of 135 B.C. as a horse comet, this may be the first recorded description of these features. However, Mithridates may have merely used the imagery of a type of comet thought at the time to predict the fall of tyrants to symbolize his struggle to evict Rome from Asia Minor!

44 B.C. (C/-43 K1), "Caesar's Comet"

This comet is widely considered to be the most famous of ancient times; however, its fame rests less on its appearance as on the circumstances of its occurrence. This was the comet that blazed in the skies of Rome following the assassination of Julius Caesar and which became immortalized by the Romans on the reverse of a coin bearing a portrait of Augustus struck in honor of the great Julius. Many centuries later, Shakespeare may also have made oblique reference to it in his play *Julius Caesar*. In the second scene of Act II, Caesar's wife, Calpurnia, warns her husband of portentous events that she interprets as ill omens for him. She tells him that during the night "Fierce fiery warriors fought upon the clouds, in ranks and squadrons and right form of war" and then continues with the famous lines "When beggars die there are no comets seen. The heavens themselves blaze forth the death of princes." Nevertheless, if these lines were inspired by Shakespeare's knowledge of the comet of 44 B.C.

and not simply by the portentous reputation of comets in general, he was either mistaken about the time of its appearance or he took artistic license with the records, as he had the comet appearing *prior* to Caesar's murder, not following it, as was actually reported.

Nevertheless, despite the comet's undoubted place in history, there are persistent problems concerning it that make its inclusion here suspect. The earliest Roman reference to it seems to have been by Caesar's adopted son Octavian (later to become Emperor Augustus) which, although not surviving in its original form, was referred to by Pliny the Elder in his *Natural History*, written around 77 A.D. According to Pliny, Octavian wrote that "On the very days of my games, a comet was visible over the course of seven days, in the northern region of the heavens. It rose at about the eleventh hour of the day and was bright and plainly seen from all lands. The common people believed that this star signified the soul of Caesar had been received amongst the spirits of the immortal gods. On this account, it was added as an adornment to the head of the statue of Caesar that I not long afterwards dedicated in the Forum."

Variations of this account were given by several Roman authors during the following centuries. For example, Seneca notes that "a comet burst forth after the death of the deified Julius, during the games of Venus Genetrix about the eleventh hour of the day." In the fourth century of our era, Servius presented an account that had the comet visible for 3 days and visible at midday and during the daytime.

Until recently, the games of Venus Genetrix were thought to have been held during late September. According to a 1999 study by John Ramsey and A. Lewis Licht, the temple of Venus Genetrix was inaugurated on September 26 in the year 46 B.C.; however, these authors also drew attention to the fact that only 2 years later, a new celebration (the Ludi Victoriae Caesaris) had been created around July 20–23 and that the earlier games had been combined with this. By 44 B.C., therefore, the games should have been held in late July. The comet, it seems, would have appeared in late July, not late September, as had previously been believed.

The significance of these dates relate to the fact that a comet, whose position in the sky and conspicuous appearance read very like the Roman one, was also recorded in China that year. Herein, however, lies the difficulty. The Chinese comet was recorded for the period of May 18 to June 16! We will take a closer look at this problem shortly.

According to Chinese records, the comet appeared at the northwest and moved into Orion. When it first appeared, its tail measured about 12 degrees but after several days grew to some 15 degrees and pointed toward the northeast. Interestingly, the Chinese records mention that the comet was "reddish-yellow in color." To a modern comet observer, this immediately implies a comet rich in dust and sodium. Large dusty comets passing close to the Sun almost inevitably show strong sodium emission lines in their spectrum, giving them a distinct yellow-orange hue. The Chinese report of the comet of 44 B.C., therefore, may well be the earliest record of sodium emission in a comet!

The Ramsey–Licht study of this comet and its association with Julius Caesar's funeral games included the first serious attempt at computing the orbit from the meager available material. According to these authors, the comet came to perihelion on May 25 at a distance of only 0.22 AU from the Sun. From this, Gary Kronk computed an absolute magnitude (H_{10}) of around 3.0, making this an intrinsically rather bright object. If these values are correct, the comet must have been close to perihelion at the time of its discovery in China and very bright. The intrinsic brightness of the comet, plus its proximity to the Sun at the time, is very consistent with strong sodium emission and an intense dust tail, as implied by the Chinese description.

According to the Ramsey–Licht orbit, the comet did not come very close to Earth, although there were two rather distant approaches of just under 1 AU, the first on May 12 as it approached perihelion and the second as it receded on August 1. These twin approaches, even though relatively remote, would have helped extend the duration of the comet's visibility.

However, we have a problem! As mentioned earlier, there is nothing reported in Roman or Greek records of a comet in May or June. And there is nothing recorded by the Chinese in late July!

When the Roman comet was thought to have appeared in late September, the discrepancy between the Chinese and Roman times seemed too great to be explained away, and historians of astronomy either followed the seventeenth century cometologist Pingre in ascribing the Chinese and Roman accounts to two separate and unrelated comets or else to wonder if the second comet was anything other than a story invented by the future Caesar Augustus for his own political purposes! The latter is not very likely, as his account was probably written within a few years of the event when the happenings surrounding the death of Julius remained relatively fresh in peoples' memories. Had he simply invented a comet or transposed one seen several months earlier, the falsification would not have gone unnoticed and would surely have been made public sooner or later by his opponents.

By shifting the comet's appearance back to late July, the Roman and Chinese observations are brought closer together, yet not so close as to coincide. In one sense, the problem may even have been exacerbated, as two very bright comets in much the same part of the sky in such quick succession, while not totally unheard of, is certainly very rare. Also, the Ramsey–Licht orbit shows that both Chinese and Roman observations fit a single orbit, strongly implying that only one comet was observed.

The non-recording of the comet in Rome during (what we may refer to as) the comet's "Chinese period" may have a straightforward explanation, one that will be well appreciated by comet observers. The weather! Or, to be more precise, atmospheric clarity.

In one of those remarkable coincidences that gladden the hearts of sooth-sayers, the time of Julius Caesar's assassination was marked by a catastrophic natural event; a major eruption of Mt. Etna on the island of Sicily. Evidence for the reality of this eruption comes not only from ancient western sources such as

Livy and Virgil, but is also corroborated by meteorological data from China and by the detection of sulfate deposits in the glacial ice of Greenland and tree ring evidence from North America indicative of a period of global cooling during the years immediately following 44 B.C. This is consistent with a high-altitude suspension of volcanic dust, lasting for several years, following a major volcanic eruption.

Apparently, the degree of obscuration following the eruption was so great that people became terrified at the faintness of the Sun, worrying that it might never again return to its normal brilliance. Clearly, this was more than a slight and passing haze!

The traditions of unusual events around the time of the Ides of March, once again immortalized in Shakespeare's play, may be grounded in meteorological (or maybe that should be tectonic) truth. In Shakespeare's rendition, Brutus, in the first scene of Act II, remarks "The exhalations whizzing in the air give so much light that I may read by them." The foul deed of Brutus and Cassius is presented against a background of thunder and lightning. Of course, to some degree this is simply dramatic effect, but maybe not totally without a foundation in the prevailing conditions at the time.

A volcanic haze sufficient to dim the Sun would surely obscure a comet – even a very bright one. We may speculate that as Etna's fires settled down in the following weeks and months, the haze began to thin over Italy, as high altitude westerly winds wafted the worst of it away. Maybe the haze drifted over China! Perhaps the Chinese lost sight of the comet in June, not because it faded from view but because atmospheric transparency took a turn for the worse.

This is all very well, but one major difficulty remains: the reported brightness of the comet during its Roman phase. Despite its second approach to Earth on the first day of August, the comet was 2 months past perihelion at the time of the Roman sighting, if the Ramsey–Licht orbit is to be trusted. Even according to Kronk's bright absolute magnitude estimate, the comet should have been no more than a faint naked-eye object by late July. What are we to make, then, of the tradition that it was not merely a bright object but one visible in full daylight?

Ramsey and Licht argue that the comet faded from its peak around the time of perihelion (i.e., its Chinese period) and would indeed have been very faint in July. However, they then propose that it experienced a major flare later in the month – coincidentally around the time of the games – when it was seen by Octavian.

Comets are certainly prone to temporary increases in brightness, and it is quite possible that such an event did occur. It is also possible that the comet faded more slowly than the inverse fourth power law assumed in Kronk's absolute magnitude determination. Each of these proposals or a combination of both could easily account for the comet becoming a relatively conspicuous naked-eye object following a period of obscurity, as Ramsey and Licht propose. The real difficulty, though, lies with the magnitude of an outburst capable of turning an ordinary naked-eye comet into something visible in full daylight.

This would require an outburst in the order of 10 magnitudes, that is to say, a brightness increase of 10,000-fold!

Brightness flares close to this range have been observed in one or two comets, but these objects have either been small and intrinsically faint or very remote and essentially inactive just prior to the outburst. A number of outbursts in the 4–5 magnitude range have also been observed over the years, with periodic comets proving more prone to this type of behavior. Most, though not all, major outbursts are of short duration, with the comet fading perceptibly after a day or thereabouts. Large and active comets do experience outbursts, but normally only of one or two magnitudes. Splitting of a comet's nucleus frequently results in a flare (possibly of several magnitudes) that may actually settle down into an enhanced absolute brightness, and it is possible that this happened to the comet of 44 B.C. A splitting of the nucleus can also release a fresh load of dust that rejuvenates the tail, making a comet not only brighter but potentially more spectacular than previously. The small perihelion passage of this comet (even though having occurred two months earlier) might give some support to thinking that a split occurred.

Nevertheless, while we can accept that it might have flared by, say, 3 or 4 magnitudes and ended up a relatively conspicuous object with a rejuvenated tail, it is extremely difficult to accept that it could have brightened to daylight visibility. It is advisable to treat apparently unique happenings such as this with great caution, unless absolutely every alternative explanation has been eliminated.

The evidence of daylight visibility apparently goes back to Octavian's statement that the comet first became visible "at the eleventh hour of the day." According to the ancient Roman system of dividing up the day, there were always 12 h of daylight and 12 h of night. The first hour of the day began at sunrise, noontime was always the sixth hour, and sunset both terminated the 12th hour of the day and began the first hour of the night. This meant that Roman hours differed in length at different times of the year and at different latitudes and that (except at the equinoxes) diurnal and nocturnal hours were of different lengths. This may seem strange to us, but it has a certain amount of logic and has served well in practical matters. For our immediate concerns, however, saying that the comet became visible at the 11th hour of the day means that it was first visible sometime during the second hour prior to sunset.

Incidentally, Servius's statement that it was seen at midday finds no support here, unless he was thinking in terms of a (very un-Roman) 24-h clock beginning at midnight. If the "eleventh hour" was translated as "11 a.m." the comet would have been visible at midday, or thereabouts, but that would be a departure from the Roman system of timekeeping. It should be noted though, that Servius gave the duration of the comet as only 3 days, also contradicting Octavian's statement (at least, as it has come down to us). Maybe Servius got his information from a dubious source that also confused different systems of recording time!

Be that as it may, it would really be helpful if Octavian's original statement had been preserved. Is it possible that he was misquoted and that what he really

said was that the comet became visible during the 11th hour of the night ... or
even simply, during the 11th hour? Could Octavian have meant that it became
visible at the "eleventh hour [of the night]" and Pliny misread it as "eleventh
hour [of the day]"? This would certainly solve the problem and would also
explain Octavian's further statement that the comet was "bright and plainly
seen by all." If it was already visible 2 h before sunset, adding that it was bright is
surely a bit redundant!

Comet 178 A.D.

Undoubtedly, there were comets between the years 44 B.C. and the beginning of
the Christian era worthy of inclusion among the greatest of the greats, but
existing records are not sufficiently detailed to distinguish them from the more
usual great and bright comets that appeared during those years.

There seems, for example, to have been a comet with an exceptionally long
tail in the year 32 B.C., but it is mentioned – and then only briefly – in just one
Chinese record. Moreover, despite noting a tail of at least 70 degrees in length,
the object is described as a sparkling star, a designation more normally reserved
for comets with no conspicuous tail. Maybe the tail was faint or there is something wrong with the record. In any case, the description of this comet is too
brief and uncertain for its inclusion here.

Accordingly, the next comet that we can identify with some confidence as
having been top grade spectacular was not observed until quite late in the
second century of our era. Unlike the last entry, this comet is neither well
known nor controversial, like the Comet of Julius Caesar.

Our knowledge of the Great Comet of 178 comes from China. The most
ancient existing account, the *Hou Han shu*, records that sometime between
August 31 and September 28 in the year 178, a broom star appeared in Virgo
and was described as being several degrees long. Broom stars (*hui* or *sao-hsing*)
were recognized by the ancient Chinese as one variety of "ominous star" and are
described in ancient texts as having a body like a star and a tail resembling a
broom. They are one of the easier classes of ominous stars to identify!

The broom star of 178 moved into that part of the sky known to Chinese
astronomers as the T'ien-Shih Enclosure – a region that includes the constellations of Hercules, Serpens, and Ophiuchus ans Aquila – and its tail extended to
a length of "5–6 chang," i.e., 50–60 feet. Depending on exactly how the Chinese
measurements of angular length are interpreted in terms of degrees of arc, the
tail reached a length of either 50–60 or 75–90 degrees, and more likely, the latter!
Either way, the tail was very long, and the comet must have been an extremely
spectacular sight in the evening sky of those distant days. The tail was also said
to have exhibited a reddish color, probably indicating that it was a dust tail and,
because it must have been bright enough to register color, presumably quite
intense. The comet was said to have remained visible for over 80 days and to

have been located within the constellation of Eridanus when it eventually faded from sight.

A low precision and (apparently) unpublished orbit was computed by J. R. Hind in the nineteenth century, giving a time of perihelion in early September (exact date not determined) at a distance of 0.5 AU. The inclination of the orbit, according to this computation, is only 18 degrees, more typical of short-period than long-period comets, although the object's spectacular appearance tells a different story!

If the orbit attributed to Hind is anywhere near correct, it seems that the comet made a close pass by Earth around mid-October. Just how close it came depends, of course, on the exact day and time of perihelion, not to mention the accuracy of the orbit in general, but it is possible that it passed just 0.05 AU from our planet. If correct, this close flyby goes a long way to explain why the comet became such a magnificent sight.

Comet 191

Very little is known about this comet, but from what we can glean from available records, it seems to have been a remarkably spectacular object fully worthy of inclusion in our list.

According to Chinese chronicles, sometime during the month beginning October 6 and ending November 4, a "comet-like banner" appeared in the constellation of Virgo. It was said to be "white" (which normally indicates bright or intense) and "over 100 degrees long." This account is repeated in Korean chronicles, but nothing more is known about this object. Either it faded quickly, the weather turned foul, or for some other reason observations ceased or records were lost.

Nevertheless, from this single description and noting both the comet's position and the immense length and apparent brightness of the tail, I. Haswgawa suspects that this may have been one of the Kreutz sungrazing comets and suggests a perihelion passage within 0.01 AU of the Sun in late September or early October 191.

Comet 252

Like the above entry, this is also an obscure object that, nevertheless, seems to have been more than normally spectacular. Also like the above comet, it is another possible sungrazer candidate.

Chinese texts record the comet as having appeared in the evening sky on March 24, to the west of Aries and with a tail extending through Orion. The tail is said to have been white and may have been as long as 80 degrees. The comet remained visible for a period of 20 days.

According to Hasegawa, if this comet was a member of the Kreutz sungrazing group, the date of its perihelion passage would have been within 4 days
either side of March 17.

Comet C/390 Q1

It has been said of great people that some are born to greatness, some achieve
greatness, and some have greatness thrust upon them. Something similar may
be said of great comets. Some are great because of their intrinsic size, some
achieve greatness by approaching dangerously close to the Sun, whereas others
have greatness thrust upon them by passing close by our planet.

The comet we will look at now is one of the latter. According to Gary Kronk,
its absolute magnitude was only 7.0. But this estimate may be a bit *too* faint,
given the comet's spectacular appearance, although it does fit well with the time
of its disappearance. Perhaps its brightness development was a little unusual. In
any event, the comet was not intrinsically bright, yet its close passage of Earth
plus favorable observing geometry turned it into one of the major objects of the
fourth century.

It was first noticed by the Chinese and Koreans on or around August 21,
when in the constellation of Gemini and described as a sparkling star. Following discovery, Chinese astronomers noted it as passing the T'ai-wei Enclosure –
a region marked by Coma Berenices, Leo, and Virgo – and into the constellation we know as Ursa Major. By September 8, it was located within Ursa
Major's Big Dipper asterism and sporting a tail 100 degrees long. It seems to
have faded quickly during September and was last observed on the 16th within
the region including the constellations of Draco, Ursa Minor, Cepheus, and
Camelopardalis.

The comet was also recorded in the Occident, although as was typical of that
era, these descriptions were not as precise as the Oriental ones.

Thus, the Roman historian Philostorgius described the comet as "a new and
strange star," noting that it became visible in the east at midnight within the
circle of the Zodiac. It was "large and bright ... not much inferior to the
morning star." The comet, according to this historian, assumed a diffuse
appearance when "a concourse of stars gathered around it like a swarm of
bees." Could he be describing a swarm of short-lived fragments breaking away
from the main mass?

Following this, the light of all these stars mingled together, and the comet
assumed the form of a huge and terrible double-edged "sword." At that time,
the star that had originally appeared was located within the handle of the sword,
and from this star a "root" of light shot up through the blade itself. This sounds
rather like a bright jet of material emanating from the nucleus, a phenomenon
displayed by a number of comets throughout the ages, including Halley during
several of its apparitions. In view of this activity, and the possible disruption

suggested a moment ago, it is possible that the comet experienced a strong flare in brightness at that time, giving some support to the earlier suggestion that its brightness behavior was irregular.

Philostorgius noted that the comet eventually began to rise and set with the morning star, before heading northward and passing into the very center of the Great Bear and fading away.

This comet was also noted by the Byzantine historian Marcellinus and described by him as a "sign ... in the sky hanging like a column and blazing for thirty days." This same historian also mentioned a comet the previous year that shone in the manner of Venus for 26 days. As there are no other reports of a comet in 389, Kronk suggests that this may actually be a misplaced reference to C/390 Q1.

According to an orbit computed by Hasegawa in 1979, the comet passed about 0.1 AU from Earth on August 18. It was still approaching the Sun at that time, reaching perihelion just inside our orbit on September 5. The motion of the comet was such that it would have effectively paced Earth, remaining relatively close to our planet for longer than is often the case for close-approach comets, which frequently whiz by and, after a brief peak of brightness, rapidly dwindle to a mere speck of light. This comet, on the other hand, stayed fairly close by for several weeks.

Comet C/400 F1

According to Chinese records, this comet was first sighted on March 18, when it was located in the Andromeda/Pisces region and sported a more or less northward pointing tail of around 35 degrees length. It moved northward, passing through Ursa Major in late March, before trekking southward in April, passing through Leo and into Virgo early in April's second week. It seems to have been lost to view sometime during the month of April.

Although the Chinese records do not necessarily suggest an exceptionally spectacular comet, there are good reasons to believe that it was indeed remarkable. For instance, several European records exist of a truly great comet in the year 400. Thus, Philostorgius recounted the appearance of a "sword-shaped star" in his work *Ecclesiasticae Historiae*, written about 25 years after the event. Another *Ecclesiastical History*, this time written by Socrates Scholasticus in the year 450, tells of the appearance of a "very large comet ... such as no-one had [previously] seen" in the year 400. According to Socrates, this comet stretched "from the sky to the ground," possibly implying a tail length of 90 degrees. The author of yet another *Ecclesiastical History*, the Roman historian and lawyer Hermias Sozomen, writing around the year 443, also describes the Great Comet of 400 in fearsome terms. According to him, the comet was of "extraordinary magnitude" and "larger than any that had previously been seen."

There is another reason, however, for thinking that this comet was exceptional. According to an orbit computed in 1979 by Hasegawa, it passed through a small perihelion distance of just 0.21 AU on February 25 and also passed by Earth at just 0.073 AU on March 31. Among known comets (excepting some tiny objects observed in recent years via the SOHO spacecraft), this one holds the record for a double bill of small perihelion and small perigee. Indeed, its nearest rival in this respect did not appear until 1996! As we shall see in a chapter 5, this more recent object provided an especially fine show and is considered by some to deserve the title "Comet of the Millennium" in view of its large apparent size and phenomenally long tail. Yet, the Great Comet of 400 came even closer than that of 1996. Moreover, its close approach occurred after perihelion, not before, as in 1996, consistent with a more intense dust tail at that time (the tail of the 1996 comet was mainly gas at the time of its closest approach, although it did release more dust following perihelion). Admittedly, Gary Kronk's estimate of 6.0 for the absolute magnitude of C/400 F1 may be a little too faint to expect an *intense* dust tail to have formed, but this would be partially alleviated by the steadily closing angle between the orbital planes of Earth and the comet throughout its period of visibility. Although the dust tail was not viewed side on, it approached this orientation ever more closely.

At the time of closest approach, the comet was located in the northern circumpolar skies under conditions not too different from those of the 1996 object. Its coma, glowing intensely at around 0 magnitude, was probably several degrees in diameter. From this great nebulous mass, we may imagine a magnificent tail extending like a broadening beam of light across tens of degrees of sky. Is it any wonder that this comet was described by contemporaries as having been larger than any previously seen?

Comet C/418 M1

Evidently, this comet was not considered as spectacular as the previous entry, as it occurred between the time of the previous one and the European records describing that object as the largest ever seen. It should also be noted in this respect that the Roman historian Philostorgius wrote about both in the same book, so the fact that he favored the first can be taken as a fairly good indication that it was the more spectacular. The later object, however, apparently had such an odd appearance that Philostorgius refused to believe that it was even a true comet!

According to this author, on July 19 in the year 418 a total eclipse of the Sun occurred a little after noonday. In the darkened sky, the stars appeared, as well as a bright "meteor" (sic) in the shape of a cone of light "which some persons in their ignorance called a comet." According to Philostorgius, the object was completely uncomet-like, having no tail as such and no star-like head. On the contrary, "it seemed like the flame of a huge lamp, subsisting by itself, with no star below it to answer to the appearance of the lamp."

Apparently, this meteor became visible in the night sky following the solar eclipse, as Philostorgius goes on to say that "Its track ... was different from that of comets. For it arose first in the east, just where the sun rises at the equinox, and then passing across the lowest star in the constellation of the Bear, crossed gradually over to the west. After measuring the whole expanse of the heavens, it at length disappeared, after it had continued its course for more than 4 months. Its apex, moreover, at one time was carried up to a high and narrow point, so that the meteor exceeded the length and shape of a cone, while at another time it returned to that particular form. Moreover, it showed to the eye a number of other prodigious appearances, which showed that it was different from common stars in nature. It began about midsummer, and continued till nearly the end of autumn ...".

Several other European texts record the appearance of a comet in the east following the total solar eclipse of that year and the Byzantine *Chronicon* notes that there appeared a "blazing star" in the east for a period of 7 months.

Chinese texts mention a sparkling star appearing in the Big Dipper on June 24, 418, and a broom star in the Coma Berenices, Leo, and Virgo region on September 15, with a northward-pointing tail more than 100 degrees in length.

Although the Roman object of July through September and the Chinese September object are widely believed to have been the same, not all astronomical historians were willing to include the object seen in June by the Chinese. Nevertheless, in 1995, Gary Kronk succeeded in linking all the observations from June to September by a single orbit. If he is correct, the initial Chinese sighting on June 24 occurred when the comet was some 2.11 AU from the Sun and 2.53 AU from Earth. Assuming that the comet – then at an elongation of 55 degrees from the Sun – was glowing at magnitude of 5.5, Kronk estimated the absolute magnitude to have been a whopping 0.2!

Is 5.5 a realistic value for the discovery magnitude of a comet by an ancient Chinese astronomer? Most astronomers think that something closer to 3.0 would by more realistic. Looking back at early observations of Halley's Comet supports the view that most comets needed to be fairly bright before they were found. Assuming that Halley's behaved in ancient times much as it has during the past couple of centuries, we can make relatively confident estimates as to its magnitude at different places in earlier appearances, and it is sometimes surprising just how bright it became before being noted. For instance, during its fifth-century return, it seems to have been brighter than second magnitude before it was recorded in the same Chinese chronicle that registered C/418 M1 in June!

Moreover, in more recent times we know that some naked-eye comets seen in Europe failed to make it to Chinese records. A good example is an intrinsically very bright but very remote comet discovered with the naked eye by Father Sarabat in France in 1729. This comet was in Delphinus, and from Sarabat's descriptions seems to have been around magnitude 5.0 or 5.5. It never ventured within 4 AU of the Sun and moved only very slowly through the conspicuous group of stars that forms the constellation of the Dolphin. Yet, Chinese records are silent about this comet.

On the other hand, amateur astronomer Andrew Pearce made an independent naked-eye discovery of the short-period comet Schwassmann–Wachmann 3 in 1995 during the time of its huge outburst in brightness. At the time, this comet was not any brighter than Sarabat's, but, unlike that earlier object, it sported a strong dust tail of at least 1 degree. It was also within the conspicuous star patterns of Scorpius, which undoubtedly helped, but it was the tail that actually drew Andrew's attention. He became aware that there was something odd in that region of the sky but is very unlikely to have noticed an extra fifth-magnitude star amid the sparkle of the Milky Way!

Similarly, although ancient Chinese constellations contain stars as faint as magnitudes 5 and even 6, transitory star-like objects are much less likely to be seen, even if they are within conspicuous asterisms such as the Dipper. The fact that the June object of 418 was described as a sparkling star and not a (small) broom star implies that it did not have an obvious tail and probably appeared more like Comet Sarabat than Schwassmann–Wachmann 3. It was probably at least third magnitude at the time and, if Kronk's orbit is correct, the true absolute magnitude of the comet computes to –2 or –2.5, one of the brightest ever recorded. On the dubious assumption that it brightened all the way to perihelion according to the inverse fourth power of its distance from the Sun, the comet would have peaked around –7! The brightness increase is far more likely to have been slower than this, but it is still likely to have rivaled or exceeded Venus in luster when close to perihelion.

It is interesting to note that Kronk finds a resemblance between the orbit he calculated for this comet and that of C/1947 VI (Honda), although the latter did not pass as close to the Sun. It is not likely that these comets are in any way associated, however, and any superficial resemblance of their orbits is most probably pure coincidence. The 1947 comet was, by the way, at least 10,000 times fainter than its predecessor of 418!

As a final thought, we might like to speculate as to how Philostorgius might have compared this object and C/400 F1 had their absolute magnitudes been swapped. M1 would still have reached around first magnitude at perihelion but would not have been well placed and may even have passed unnoticed. F1, on the other hand, would have cast shadows, and it may have been possible to read by its light. By an odd coincidence, history repeated in 1996 and 1997 when the comet most closely resembling C/400 F1 was followed by a much more distant one of great intrinsic brightness, but that is a story for a later chapter!

Comet 467

The phenomenon of 467, as noted in the Chinese text *Nan Chi'I shu*, could have been either a great comet or a magnificent auroral display. Nevertheless, two European texts, the Byzantine *Chronicon Paschale* and the *Chronographia* of Theophanes the Confessor, mention luminous celestial phenomena occurring around the same time which, taken together, support the view that a very spectacular comet appeared that year, even though virtually nothing is known about it.

According to the Chinese text, a "white vapor" was seen on February 6, which was said to have stretched half-way across the heavens from southwest to southeast. This sky location, by the way, reads more like a comet than an aurora, which we would expect to concentrate in the north of the sky from a northern hemisphere site. The vapor was described as a "chhang-keng" or "long path," a type of ominous star known to the ancient Chinese astrologers and described by them as being like a roll of cloth extending across the heavens. It is depicted as having the appearance of a comet with a nucleus and two tails.

The Byzantine text says that in that year there "appeared in the heavens a very great sign, called by some a trumpet, by some a spear, and by some a beam." It was said to have remained visible for "some days," once again seeming more descriptive of a comet than an auroral display.

Finally, Theophanes recounts that "a sign appeared in the sky, a cloud in the shape of a trumpet, each evening for forty days." Although he related this to the years 465 or 466 (more probably the latter), he did not write until 813, so the event was already distant history in his own lifetime. The similarity with the above Byzantine account makes it most likely that the same event is intended, and the long duration, once more, strongly indicates a comet. Lacking evidence to the contrary, we believe that the phenomenon of 467 was indeed a comet. Moreover, from the admittedly scant descriptions available, it seems to have been one of unusual size and splendor, easily deserving of inclusion in any list of the greatest of the greats!

Interestingly, in 2007, Z. Sekanina and P. Chodas published a paper as part of their ongoing study of the Kreutz sungrazing group of comets, in which they argue that this may be a previously unrecognized member of this family and a good candidate for the parent of many sungrazers of more recent centuries. More will be said about this in Chapter 6.

Comet X/676 P1

Although no orbit for this comet has been computed, it must have been a very spectacular object, and its long period of visibility, spanning nearly 3 months, indicates that it was probably an intrinsically large object as well. Descriptions of the comet are found in texts from China, Japan, Korea, Italy, Ireland, England, Scotland, and Syria, testifying to the impression that it made upon the population of the time.

The oldest surviving account, written within 50 years of the comet's appearance, is from Japan and states that "A star appeared in the East" sometime between July 16 and August 14 of 676. It was described as being about 10–12 degrees long initially, but during the period between September 13 and October 10, it was said to have "stretched across the sky." (An alternate reading says "at length disappeared from the heavens." However, as other records say that it peaked during the September/October period, the former version would appear to be the more accurate.)

The English historian, the venerable Bede, records that in the year 678 "there appeared in August a star which is called a comet; and continuing for three months, it rose in the morning, sending out, as it were, a tall pillar of flame." Bede borrowed from an older source, and it is accepted that the events which he ascribes to 678 actually occurred 2 years earlier.

The comet was first seen in China on September 3 in the constellation of the Twins. The tail was then about 4 or 5 degrees long, eventually growing to 45 degrees or thereabouts. The comet, according to Chinese sources, remained visible for 58 days, that is, until around November 1.

Korean records also tell of a comet in Gemini during August and September. Similarly, the *Anglo-Saxon Chronicle* records that "the star comet appeared in August, and every morning for three months shone like a sunbeam."

In Italy, the comet was described as having "very brilliant rays," and as lasting "for three months," while the Irish *Annals of Tigernach* simply record that "a bright comet was seen in the months of September and October." A Syrian record by Agapius of Manbij, though, dated the following year, almost certainly describes the same comet. According to this author, "An awesome comet appeared every morning from 28 August to 26 October, sixty days in all."

Comet C/770 K1

This broom star first became visible in the morning sky from China on May 25 as a white (implying bright or intense) object with a tail that, depending upon how the Chinese measurement is interpreted, was either 50 or 75 degrees long. The head of this magnificent object was then located in the Auriga/Taurus region, a mere 18 degrees from the Sun. The comet continued to move east, traversing Canes Venatici around July 9 and remaining visible until the 25th of that month, when it was last recorded in the evening sky.

In Korea, the comet was apparently noticed first on the morning of June 9 and in Japan, sometime during the period of June 28 to August 25.

An orbit computed by Hasegawa in 1979 indicates a perihelion of 0.58 AU on June 5 and a minimum approach to Earth of just 0.3 AU on July 10. The absolute magnitude was probably around 3, making this an intrinsically rather bright comet that probably showed an intense dust tail. This tail would have been seen edge on about the time the comet was passing closest to earth, making for a very spectacular sight indeed!

Comet X/838 V1

This spectacular comet was first seen in China on November 10 as a broom star some 30–40 degrees long, apparently in or near the constellation of Corvus. Two days later, it was spotted by the Japanese, who located it in the southeast and in one record described it as being "a star like a shining sword." By November 13, the tail seems to have exceeded 50 degrees, according to Chinese records.

The Japanese monk, Ennin, wrote in his diary that "From midnight until dawn I left my room to look at the comet. It was in the southeastern corner, and its tail pointed to the west, shining very distinctly. Seen from a great distance, the length of its tail might be estimated at over a hundred feet. People all said that it was definitely a shining sword."

It may be assuming too much to simply translate "100 feet" as "100 degrees" in this context, but a further Chinese record on November 21 suggests that this is not impossible. The tail was then said to have "[stretched] across the heavens from east to west."

The comet seems then to have disappeared in the Sun's rays. Chinese records make a further statement that it "went out of sight" on December 28, but as there is no mention of the comet between late November and that date, it is probably better to follow Gary Kronk's suggestion that this statement only means that it was looked for, but not found, on that date.

In any case, sources from China, Japan, Germany, and Belgium tell of a great comet in January and February of the following year. This comet (termed "terrible" in the German records) seems to have been very bright and, according to the Japanese, sported at least 15 degrees of tail. It appeared in Chinese annals on March 12 and continued until April 13 or thereabouts.

There is some difference of opinion as to whether the great January comet and the March/April object refer to the same comet; or even if the latter was a comet at all (His Tse-Tsung was of the opinion that the March object was a nova). The bigger issue, however, is whether the object(s) seen in January and March can be identified with X/838 V1. Ho Peng Yoke considered all three sets of observations to relate to the one comet, X/838 V1, but Kronk doubts this identity in view of the considerable gap in observations during December. Except for the ambiguous note about the invisibility of the comet on December 28, nothing is said about 838 V1 during the period from late November until nearly the end of January, the following year. This would appear to cast a good deal of doubt on this comet's supposed identity with the January object.

Considering that Halley's Comet had only recently (in February 837) made an unusually spectacular return during which it approached earth closer than at any other time in recorded history, the late 830s present themselves as an exciting time for great comets, although the people of the time may not have been so sanguine in view of the fear with which these objects were then held!

Comet X/891 J1

On May 12, 891, the Chinese recorded the appearance of a broom star in the constellation of Ursa Major. Between that date and the time the comet went out of sight on July 5, it trekked eastward through the "T'ai-wei Enclosure" (the region of sky marked by Coma Berenices, Leo, and Virgo), swept Alpha Bootes and the region comprising the constellations of Hercules, Serpens, Ophiuchus, and Aquila.

Unfortunately, the descriptions of its path are not accurate enough for an orbit to be calculated, but the comet was obviously a very spectacular one, judging by the remark that it was "over 100 feet [presumably degrees]" long. This great tail length probably implies a rather close approach to Earth sometime during the comet's period of visibility.

The comet can also be identified in the chronicles of other cultures. It is possibly identical with the "guest star," sometimes thought to have been a nova, recorded by the Japanese on May 11, but is given a more positive description in several European records compiled during the following centuries.

Thus, the Winchester edition of the *Anglo-Saxon Chronicle* (dating from 1154) says that "[in 891] after Easter [April 4], during the Rogation days or earlier, appeared the star which in Latin is called *cometa*; some men say in English that it is a 'haired star', because a long ray stands out from it, sometimes on one side, sometimes on every side." Other European texts simply state that a comet and an eclipse of the Sun occurred in the year 891. The eclipse, by the way, took place on August 8, and there seems to be no connection made between the two events.

Finally, Islamic texts dating from the thirteenth century note that in the year 891, there arose "a very brilliant star" on May 13. Afterwards, according to these texts, "its brilliance became locks of hair."

Comet 893

It seems that just 2 years after the above-mentioned object, another comet of prodigious size blazed across the heavens. Once again, however, the records are not very precise, and an orbit cannot be computed. Indeed, less is known about this one than the above, even though it must have been a truly remarkable phenomenon.

The Chinese astronomers of the day apparently first saw the comet on May 6, following several weeks of cloudy weather. As the clouds dispersed on the evening of May 6, the great comet became visible in the constellation of Ursa Major and was reported as having a tail 100 degrees long. Passing along a path that apparently mimicked that of the 891 comet, this one was said to have extended its tail to about 200 degrees and remained visible for 37 days until cloudy weather again closed in and prevented further observation.

The length of 200 degrees seems excessive. Did its tail really extend further than from horizon to horizon?

A more serious question though, relates to the similarity between this comet and the previous one, not just in appearance and size but also in time of visibility and apparent path through the constellations. Is it possible that the two comets were really one and that the records for 893 actually should refer to 891?

Although this remains a possibility, it seems strange that the same chronicle, *Hsin T'and shu,* details both comets. If the record of the second comet appeared in a different chronicle, we might have cause to become suspicious, but for the same chronicle to double up like this is a lot less likely.

Also, the dates and descriptions are not *exactly* the same. Whereas the second account does have a lot of similarities with the first, it is not simply a word-for-word copy of it. This gives it a certain ring of truth.

Yet another and more interesting possibility is raised by this. Could the two comets have been pursuing the same, or very similar, orbits? Is this an instance of a genuine comet pair, the product of a nucleus that split during a previous return, or as the comet drifted inward toward its ninth-century perihelion?

This is a possibility that should be taken seriously, although it is unlikely that any more light will be shed upon it unless some currently undiscovered records turn up in the future.

Comet C/905 K1

This comet was first sighted, again by Chinese astronomers, on the evening of May 18 and was then said to have resembled Venus, though emitting rays like a broom star and measuring some 50 degrees or more in length. It was located in the northwestern evening sky and was described as being "blood red" in color. The next evening, though, its color was said to have changed to that of "white silk." We wonder if the evening sky on May 18 was covered with a thin veil of cirrus cloud or dust that may have given it an unusual ruddy color. If that is the explanation, it says something about the brightness of the comet and its tail to shine through a veil in that way.

On May 19, the comet was also seen in Iraq and recorded as being located in "the far part of Pisces" and "rising" (presumably in the sense of becoming visible) at the time of evening prayer. By May 21, the Japanese were also noting the broom star in the evening sky.

Both the Chinese and the Japanese astronomers were especially impressed by the length of the comet's tail. On the 22nd the former recorded that it was at least 45 degrees long. The head was located near the "Twins" of Gemini, but the tail penetrated into Ursa Major. On the 30th of the month, the Japanese claimed that the tail was 300 degrees (sic.) long and that it "extended across the heavens" on the following evening. The "300 degrees" is almost certainly either a mistake or hyperbole, but the remark about the tail's length on May 31 suggests that it was, nevertheless, very long.

According to the Japanese, the comet dimmed quickly in early June and went out of sight on June 8. This is not in accord with the Chinese report, however, and may have more to say about the weather conditions in Japan at the time than about the behavior of the comet. According to the Chinese, the comet was still very intense on June 12. In words reminiscent of the Japanese description 12 days earlier, they recorded that on this evening the tail "stretched across the heavens" from the region near Leo and Lynx almost to Aquila! It would appear from these descriptions that the tail of this comet was exceptional in both its length and intensity.

The nights, following this observation, were overcast. When clear skies returned on June 18, the comet had disappeared.

This comet was apparently seen in Europe, but the records from there are vague, and little additional information is added. One account telling of "a brilliant comet which cast out its rays toward the east and remained conspicuous for 40 nights" probably refers to this object, and another records a comet "in Pentecost" (May 19 in 905), but gives no further details. A French text speaks about a comet "discovered in the north" later crossing the Zodiac "between Leo and Gemini" and remaining visible for 23 days, but no physical description is given. Nevertheless, the very fact that even brief mention of the comet found its way into the annals of so wide a variety of lands and cultures itself speaks of the splendor of this grand object.

I. Hasegawa is the only astronomer who has attempted to compute an orbit for this object (in 1979), and his calculations suggest a perihelion passage on April 26 at just 0.2 AU from the Sun. Based on this orbit and available descriptions of the comet, Gary Kronk estimates an absolute magnitude of 4.5. At the time of its May 18 discovery from China, the comet would have been emerging from twilight at 21 degrees from the Sun, just 0.32 AU from Earth and 0.73 AU from the Sun. Based on Kronk's absolute magnitude estimate, it would have been about magnitude 0.6. The Hasegawa orbit indicates a minimum approach to Earth of just 0.198 AU on May 25, according well with both the comet's brilliance (first magnitude) and the extreme length of tail visible then. At that time near the plane of the comet's orbit, Earth's inhabitants viewed the dust tail edge-on, accentuating both its length and intensity.

With the Great Comet of 905, we reach the last of our list of the greatest of the greats from the earliest times until the end of the first millennium of our era. Undoubtedly there were others that have slipped through our net. These ancient times necessarily lack records of sightings from the southern hemisphere, for instance, and surely there must have been very great comets passing through far southern skies that have left us no record. Maybe some of the normal comets that we passed over became "extraordinary" after passing from the ken of the northern astronomers. We simply do not know.

In any case, we have now arrived at the start of the millennium during which the chief mysteries surrounding comets found their solution. During the next millennium, comets came to be accepted as bona fide astronomical bodies, a change in perspective brought about by the painstaking observation of a large number of objects, some of which were truly outstanding members of their class. We will look at these in Chapters 4 and 5, beginning with the first comet of the second millennium that seemed to be an extraordinary spectacle (apart from Halley's in 1066, which we have already discussed) and ending with the first extraordinary comet of the third millennium, still faintly visible as these words are being written.

Truly, the story of the greatest of great comets weaves a thread throughout the history of our race!

Chapter 4
The Greatest Comets from A.D. 1000 to 1800

We now come to a time when the recording of bright comets improved, and fewer truly grand objects are likely to have slipped through the net of preserved chronicles. Still, we must admit that our list is not complete, especially for the earlier centuries. The first comet that seems clearly to meet our criteria for inclusion in this list is the magnificent apparition of Halley in 1066, but we have already spoken of this in Chapter 2.

Another comet which may have become a member of the select group is the comet of 1075. No orbit for this object has been calculated, and it does not even bear an official designation; yet it is possible that it belongs to the Kreutz Sungrazing group and more about this will be said later. None of the surviving records of the comet imply anything exceptional, but if it was a sungrazer, the available observations would be confined to the pre-perihelion leg of its orbit, when the tail was still developing. However, sungrazers appearing at that time of year (November) reserve their best performance for southern latitudes *following* perihelion. If the 1075 comet was really a member of this select group, it would undoubtedly have been magnificent in the southern hemisphere during late November and December where, however, no records were being kept!

Accordingly, the first comet of the second millennium of interest to the present chapter is the widely observed object of 1106. This is also suspected of being a sungrazer, as we shall see, but it appeared at a time of year when both northern and southern hemispheres could behold its splendor!

Comet X/1106 C1

This comet seems to have been seen first in Belgium during broad daylight on February 2, 1106. According to historian Sigebertus Gemblacensis, a star appeared about 1 degree from the Sun between the third and ninth hours. He also noted that on February 12, near Bari in Italy, a star was likewise seen during the day. Is it possible that the "12" is actually a misprint for "2," making these observations on the same day? It seems unlikely that the comet would still (again?) be visible in daylight 10 days later.

D. Seargent, *The Greatest Comets in History*,
DOI 10.1007/978-0-387-09513-4_4, © Springer Science+Business Media, LLC 2009

Be that as it may, the next recorded sighting of the comet was from Palestine on February 7, when it appeared in the western sky after sunset. This observation locates the comet in Pisces, with a tail of around 100 degrees extending all the way back into Gemini.

Two days later the Japanese spied it, also estimating the tail as around 100 degrees in length and "white" in color. By the 11th, according to the Japanese astronomers, this tail had shrunk to approximately 15 degrees, and the next day, to a mere 5 degrees. However, by February 20, it was back to 30 degrees or thereabouts. Presumably, weather conditions were variable in Japan that winter! After more than 30 days, the comet faded from sight.

Chinese and Korean astronomers also noted this comet. According to the latter, it became visible on February 9 and had a tail of about 15 degrees in length. It was located in the southwest and stayed visible for over a month.

The comet was first recorded in China on February 10 and described as a "broom star" with a head "the size of a cup" located in the western sky. The tail, estimated as 90 degrees in length and 4–5 degrees in width, pointed obliquely toward the northeast. This last remark agrees with the Korean observation locating the comet in the southwest. The Chinese location "in the western sky" is probably best taken as a general location only. The Chinese also added the curious statement that the "rays" of the comet were "scattered in all directions as if they were broken up into fragments." It is not easy to understand what they meant by this, but (with the tail of Comet McNaught still fresh in memory) we wonder if the tail of the 1106 comet was "broken up" by a spectacular series of striae similar to that displayed by this recent object. More will be said about this possibility later.

Records of this comet throughout the month of February are found in the annals of cultures from all over the then-known world. An Armenian text speaks of "an awful, big and amazing comet, which frightened those who saw it" appearing on February 13 and remaining visible for 50 days. Its tail is said to have covered "most of the sky" and to have appeared "like a flowing river." Nobody, according to this text, "had heard of such a strange phenomenon."

By February 16, the comet was being recorded in the monastic histories of England, Scotland, France, Italy, The Netherlands, Belgium, and Germany.

A Welsh chronicle describes it as "a star wonderful to behold, throwing out behind it a beam of light of the thickness of a pillar in size and of exceeding brightness."

In England, the comet was seen on the evening of February 16 as "a strange star in the south and west." It was said to have shone for 21 days in the "same form and same hour" and, though in itself "small and dim" nevertheless possessed a tail that was "exceedingly clear; and flashes of light, like bright beams, darted into the star itself from the east and north." The record then adds the puzzling comment that "Many affirmed that they saw several strange stars at this time."

This latter comment is interesting. It may mean nothing more than the purely coincidental occurrence of a shower of meteors, or even a few random bright ones, at the time of the comet, but there is another and more interesting possibility. We have already mentioned that this comet may be a member of the Sungrazing group. As we will see in Chapter 6, the great Sungrazing comet of 1882 was found to be accompanied for a short while by a number of comet-like formations located at some distance from the main object and having the appearance of satellite comets. At least one of these was actually reported as a newly discovered comet. None of the 1882 objects seemed to have reached naked-eye visibility, but it is interesting to speculate that, perhaps, similar formations attended the comet of 1106 and, just maybe, became bright enough to be spied by the ancient English sky watchers of that year.

But what are we to say about the "flashes" and "bright beams"?

At face value, these appear to have happened too rapidly to be explicable by any known process taking place within the comet, and we might be inclined to explain them away as something atmospheric. But what if these "flashes" were not quick events? Maybe they were structures within the tail. Could this be another reference to the "rays" mentioned by the Chinese? Again, we suggest that striae within the tail may be meant here. If this suggestion is correct, we may have here the first references to such details in a comet's tail.

This interpretation, perhaps, is strengthened still further by a remark in the Peterborough *Anglo-Saxon Chronicle* (1154) that states, "One evening it was seen as if the beam were forking in the opposite direction, toward the star." Because the "beam forking in the opposite direction" was said to have been directed "toward the star," it seems not have been an anti-tail, as might be supposed by a quick reading of this account. It seems more like a feature within the tail itself – a bright striation perhaps?

The comet was also recorded in Muslim lands. The text, *al-Kamil fi alta'rikh* of 1233 records that sometime during the period from December 1105 to January 1106, "a star appeared in the heavens with locks of hair like a rainbow. It went from the west to the middle of the heavens and it was seen close to the Sun before its appearance during the night. It continued appearing a number of nights, then it disappeared." A similar account in a text dated 47 years earlier places the comet in the 1106 February–March period, in conformity with the accounts from other parts of the world.

Although an orbit has never been calculated for this comet, astronomers have long been interested in its association with more recent objects. Thus, Halley thought it to have been a previous apparition of the Great Comet of 1680, as we saw earlier. This proposed identification failed to hold up to scrutiny.

More interesting is the possible (maybe even probable) association with the Kreutz Group of Sungrazing comets and, in particular, with the brilliant comets of 1882 and 1965. We will look at this more closely in Chapter 6, but it is worth mentioning here that those who were around to see Ikeya-Seki in 1965 can relate to the 1106 descriptions of a "small and dark" or "small and dim" head from which sprouted a magnificent tail that "was very bright and seemed like an enormous beam." The comparison of the comet's tail with a rainbow in the Muslim text quoted above also conjures up a sense of déjà vu. Alan Marks, then Director of the Comet Section of the Astronomical Society of New South

Wales, described the tail of Ikeya-Seki as "being like a rainbow," the exact phrase used 732 years earlier to describe the comet of 1106. (Of course, this says nothing of the color of the tail, nor even that the tail was shaped like an arc. It simply means that it swept across the sky in a relatively narrow band of fairly uniform intensity.) As we shall see, this description well fits the tail of a great sungrazer, and the similarities of description between 1106 and Ikeya-Seki may be much more than coincidence. But more of that in Chapter 6!

Comet C/1132 T1

Nobody at that time knew it, but late on August 30 in the year 1132, a large and potentially bright comet passed through perihelion 0.74 AU from the Sun. Having approached from beyond the Sun, it remained deep in twilight and unnoticed as it headed almost straight for our planet!

Not until October 3 did the comet become sufficiently conspicuous to attract the attention of medieval astronomers. Once again, it was the Chinese who first spied it, recording it as a broom star in the constellation of Ursa Major.

The very next night, the Japanese and Koreans also saw the broom star and noted that its tail pointed toward the southwest. According to both groups of astronomers, the comet was visible all night long, albeit most obvious in the early morning hours.

By October 7, it had moved to Aries and became widely observed in both Orient and Occident. According to one Japanese text, the tail had lengthened to around 45 degrees, and the comet stayed visible all night; rising in the east at sunset and setting in the west at sunrise.

Following its peak display on the 7th, the comet faded quite rapidly and the tail grew shorter. It had already shrunk to around 15 degrees on October 8 and was down to only 3 or 4 degrees the following night. By the latter date, the comet was favoring evening skies, although it still could be followed for most of the night.

Japanese astronomers were clouded out for the next two nights, but managed to see the comet (apparently for the last time) on October 12. The Chinese did better and tracked it until October 27.

According to an orbit computed by S. Ogura in 1917, C/1132 T1 should have passed Earth at a mere 0.045 AU on October 6. Assuming that the comet was followed by Chinese astronomers until it faded to around magnitude 5 and that its brightness development could at least roughly be described by the inverse fourth-power law, Gary Kronk estimated its absolute magnitude at about 4.5. This implies a maximum brightness in the neighborhood of −2.2, similar to that of the planet Jupiter, at the time of closest approach to Earth. Visible all night, almost opposite the Sun in the skies, the comet must have been a truly magnificent sight for a few nights around that time. Under dark and unpolluted medieval skies, this comet may have been capable of casting weak fuzzy shadows.

Comet C/1147 A1

If the previous comet became numbered among the greatest of the greats chiefly because of its close approach to Earth, C/1147 A1 achieved this status principally because of a close approach to the Sun. According to an orbit computed by I. Hasegawa in 1979, this comet passed just 0.12 AU from our star on January 28, 1147!

Prior to Hasegawa's study, most astronomical historians thought that two bright comets had appeared in early 1147. The first comet was discovered by the Chinese on January 6 (more recently corrected to December 29 of the previous year) and remained a striking sight until lost in the evening twilight sometime between January 14 and 20.

The second comet was first noted, again by the Chinese, in the morning sky of February 7 and was last seen by the Japanese on February 24.

These "two" comets moved in opposite directions, the first heading southward and the second, north. According to Hasegawa's calculations, however, both sets of observations can adequately be explained by a single comet that moved in toward the Sun from a northerly direction, made a hairpin bend at very small distance on January 28, and headed back in the opposite direction. Although it probably would have been bright enough to see in daylight around the time of perihelion, apparently no such observations were made, and a period of invisibility of more than 20 days centered on perihelion only added to the impression that two separate objects were involved.

Although the Chinese were the first to see the comet, their early account of it was brief and amounted to little more than a simple statement that a comet appeared in the region of Pegasus and Aquarius. However, they did describe it as a broom star around the time of the New Year, implying that a curving tail was already visible.

It is to the Japanese that we owe the most comprehensive descriptions of the pre-perihelic leg of this comet's orbit. They apparently saw it first on January 4 and, like the Chinese, classified it as a broom star. They added, however, that the tail was between 30 and 45 degrees long on that evening. It would seem that this tail was probably rather faint over much of its length, as the estimates of its dimensions varied greatly on the following evenings. Thus, on the evening of the 6th, they estimated it as just 15 degrees but recorded an amazing 100 degrees just three nights later!

This immense tail length on January 9 is the chief reason for including the Great Comet of 1147 in our list. Even *if* the tail was not especially bright over much of its length, its dimensions alone must have made it a truly spectacular sight in the evening skies.

The comet had already passed its closest to Earth (0.32 AU) on December 29, 1146, making the inward leg of its orbit the more favorable as seen from our planet. On the way out, it moved on the other side of the Sun, and its second appearance in February was therefore not as brilliant or as spectacular as its former one, although there would probably have been the compensatory factor of an intrinsically larger tail following the comet's very close encounter with the Sun.

Once again, it was the Chinese who first spied the comet on its emergence into the morning skies but, once again also, it was the Japanese who left us the most comprehensive descriptions. According to the Chinese reports, it was sighted on February 7 sporting a tail of around 15 degrees. The following morning, the Japanese saw it and classified it as a "sparkling star," but by the 12th they were describing it as a broom star in the Aquarius/Equuleus region with a tail estimated as 15 degrees long. The comet moved gradually northward and faded, going out of sight around February 24.

Although there are no clear records of this comet in the annals of other cultures, it is possible that it was the one said to have preceded the April departure for Asia Minor of King Conrad III of Germany and Louis VII of France at the start of the Second Crusade. No precise date or description of this alleged comet is given, but the Great Comet of 1147 seems the most promising candidate.

Comet C/1264 N1

Our next comet on the list was one of the most spectacular and widely observed of the medieval period, with references to it spread throughout chronicles of both Oriental and Occidental cultures.

By contrast with most of the comets we have looked at thus far, the first sighting of the Great Comet of 1264 was made in Europe. According to Thierri de Vaucouleurs, writing the following year, it was seen in Gaul on July 17, some 4 days before it first entered the chronicles of Japan. Unfortunately, no further details of this observation are forthcoming.

In Japan, the comet was described as being in the northwestern evening sky, initially within the constellation of Ursa Minor, but moving into Leo by July 23 and Cancer 2 days later. At that time, it was again noted in European records.

It seems to have escaped notice for a brief period following these early evening observations but was picked up as a spectacular early morning object by Chinese astronomers from July 25. It was then that the first physical descriptions of the comet started appearing in the records, with the Chinese describing it as sporting an immense tail measuring over 100 degrees and "[illuminating] the heavens."

The Chinese continued their description of the comet with the interesting statement that "it became invisible only when the Sun was high up" and that "This lasted for more than one month." By this, they obviously did not mean the comet itself, which was later said to have "lasted four months before it finally went out of sight." They must have meant that for a period of more than 1 month, the comet remained visible until the "Sun was high up"!

Taking these words at their face value seems to imply that the comet remained visible in daylight (at least, for a time after sunrise) for a period of over 1 month! Such an extended time of naked-eye daylight visibility is unprecedented in the history of recorded comet observation.

Although the orbit of this comet is not well established, the one computed by Martinus Hoek in 1857 accounts for the observations quite satisfactorily, but makes extended daylight visibility – or daylight visibility at all for that matter – unlikely. This orbit gives a perihelion distance of 0.82 AU on July 20, about the time of the first observations in the evening sky. It also indicates an approach to Earth of just 0.18 AU on July 29, around the time that Chinese astronomers were noting the extremely long tail. On the assumption that this orbit is at least fairly close to the truth, S. K. Vsekhsvyatskij found that the reported duration of the comet's visibility could be accounted for by an absolute magnitude in the 3–4 range.

Assuming an absolute magnitude of 3, the comet would have reached an apparent brightness of around –1, and remained a bright object for an extended period. A comet of magnitude –1 with a 100 degree tail must surely have been a wonderful sight to behold in the morning sky, but it is hardly likely to have remained visible to naked eyes following sunrise. Indeed, the comet's very proximity to our planet at that time means that its light was distributed over a large coma diameter and therefore lacked the intensity of brilliance expected for a comet of equal brightness at small solar distance. Comets of the latter type have been seen telescopically in the daytime when at similar or somewhat fainter magnitudes, but –3 is probably the bare minimum brightness required for naked-eye visibility relatively low in the sky with the Sun having just risen. It is not likely that the Great Comet of 1264 became as bright as that, let alone that it maintained this brightness for over a month!

We suspect that "the Sun being high up" is more likely to refer to the portion of sky being brightened by the Sun reaching high up into the heavens, not the Sun itself being well over the horizon. In other words, we suspect that it means the comet remained visible until well into morning twilight, until the sky was extensively lightened by the encroaching dawn. This fits more easily with what we think we know about this object, but even so it tells us much about the continuing high luminosity of the great comet.

Chinese records note that the comet remained visible for 4 months and that "[by September 21] its rays slightly decreased." It is not clear whether this means that the tail remained close to 100 degrees long from late July until mid September or simply that the comet's brightness (and probably the intensity of the tail) had slightly faded by then. Although the latter may seem the more reasonable interpretation, there is evidence that the tail was very long in early September and may even have decreased for a time before increasing again, but more about this in a moment.

Outside China, tail-length measurements often show discrepancies, possibly reflecting meteorological conditions or other local circumstances. Thus, the Koreans measured only about 11 degrees of tail in late July, but they added the interesting fact that it was multiple; divided into five separate branches! In

Japan, the tail was measured as just 4 or 5 degrees on July 27, but was said to have "extended across the heavens" a mere 3 days later.

The Koreans, too, observed a dramatic increase in the tail's length, but they placed the maximum length around September 11, by which time they also wrote that it "extended across the heavens." Interestingly, prior to its achieving this maximum length, Korean observers remarked that the five branches noted earlier "reunited" into a single tail, which then increased in length. This was said to have happened around August 22, which is close to the time that Earth passed through the comet's orbital plane, according to the Hoek orbit. The "closing" of the tail and its increased length is just what we should expect to see during the passage through the orbital plane of a large and dusty comet.

It seems that the tail was very long in late July principally because of the comet's close passage by Earth. At that time, it was probably quite broad and fan-shaped. Maybe the bulk of the tail was relatively faint then, which could account for the discrepancy in length estimates between different locations and, indeed, on different nights as is apparent in the Japanese accounts. The five "tails" recorded in Korea may have been brighter features embedded within the broader, fainter sheath of the general tail. Although pure speculation, these features might have been synchronic bands similar to those displayed so beautifully by Comet West in 1976, or they may even have been striae. From the brief Korean description, however, it seems they were true synchronic structures. As Earth moved closer to the plane of the comet's orbit, these bands would appear to have closed together, just as the Koreans recorded.

Moreover, Earth's vantage point near the comet's orbital plane meant that in late August and early September, the broad dust tail was being viewed more or less edge on and, as mentioned above, this would have both intensified its light and increased its apparent length. Once again, this fits nicely with the Korean records.

Interestingly, the multiple structure of the tail is also implied by an Islamic text *Suluk li-ma 'rifat duwal al-muluk*. This speaks of a "star with a tail" appearing in the east in late July of 1264, rising just before dawn. The "star" was said to have given off "immense rays in the atmosphere just before its emergence," and then the interesting remark is made that "it would appear with lighted streaks like elevated fingers in the atmosphere of the heavens in the [north-west] as well as after the late evening for many nights." From this description, it would seem that the multiple "tails" recorded in Korea were here seen projecting above the horizon even though the comet itself had not at that time risen. Indeed, it seems that these "fingers" of light remained visible all night as the comet swung below the northern horizon toward its place of rising in the early dawn. What we may have here is an ancient incidence of a similar appearance to that displayed by the Great Comet of 1744 (see below) and much more recently by Comet McNaught in January 2007, as we will see in the next chapter.

The comet is recorded in many other chronicles, but for the most part little extra information is given. British texts simply state that "a comet ... before daybreak in the month of August" was observed, although some misdate the event as the following year. Likewise, a French record places "a terrifying comet ... directing a beam toward the east" in the year 1266, almost certainly a misplaced reference to the 1264 comet, and a German text has a comet "remaining visible for three months" in 1263!

As an interesting aside to the story of the Great Comet of 1264, we should mention that there was a time when a number of astronomers thought that the comet had returned as the object now known as C/1556 D1. Apparently, this idea got its start when, in 1751, Richard Dunthorne first computed a possible orbit for the 1264 comet. Probably influenced by the possibility that the comet of 1682 might return within the following decade, Dunthorne compared his newly derived orbit for 1264 with other published comet orbits and was struck by the similarity with the one that Halley had computed for the comet of 1556. Although less than convinced that he had found another periodic comet, he noted that if the two objects were in fact identical, the next return should occur around 1848. Dunthorne was supported in this by cometologist A. Pingre, whose own calculations of the 1264 orbit convinced him that this was indeed an earlier return of 1556.

Alas, the year 1848 came and went without the comet's return, but even this did not completely settle the issue. Indeed, a rather acrimonious debate developed between astronomers J. R. Hind, who still maintained the two comets to be one and the same, and M. Hoek, who did not. These two astronomers exchanged a number of intellectual blows over the matter in 1856 and 1857, before a third party entered the fray in the person of B. Valz. Valz computed three possible orbits for 1264 N1, but concluded that the uncertainties involved in each were simply too great to come down definitively on one side or the other. The comet of 1264 might have been an earlier return of 1556 ... or it might not. The evidence was just too uncertain.

In the final analysis, it seems that Hoek was correct after all. No comet observed during the past two centuries displayed any orbital similarities with either 1264 or 1556.

Comet C/1402 D1

The fifteenth century was ushered in by a comet that must surely be rated as one of the very greatest of the greats. Not only was it visible for at least 2 months, sporting a magnificent tail that spanned one quarter of the vault of the heavens, but it became visible in broad daylight for a full 8 days, a record for extended naked-eye daylight visibility not equaled to this very day.

Although one European record mentions that a "long-tailed star" became visible on January 3, most chroniclers date the first appearance of the comet in early February. The most detailed account comes from Jacobus Angelus, who states in his *Tractatus de Cometis* that "around the beginning of February a comet appeared ... for many days." By March 15, according to this author, the tail was about 45 degrees long and "shaped like a pyramid measured linearly." The apex of the pyramid was the head of the comet and the base "continued toward the upper part, diffusing through the distant side of the figure." The description implies a beautiful spreading fan, about which Angelus commented that "I never saw such a bright and colorful tail."

Continuing his account, Angelus noted that on March 22, "in the second hour, [the comet] was seen near the Sun at a distance of [31 degrees] to the north" and that on March 26/27, "in the east before sunrise its vestiges appeared, because I saw three long and very thick hairs, and after sunset I saw one hair in the west." That was to be Angelus' final view of the comet.

The mention of "three long and very thick hairs" is interesting and suggests either synchrone bands or striae within the tail. The appearance of a single "hair" in the western sky following sunset may indicate a strong curvature of the dust tail, or even the appearance of an ion tail widely separated from the fan of dust. The broad fan shape that Angelus mentions indicates that the dust tail was being viewed essentially face on at the time, which is consistent with a wide separation between this and any straight Type I tail that the comet may have possessed. As this was, clearly, a very large comet, a conspicuous ion tail very likely existed.

Strangely, the Chinese were silent about this comet, although it was observed in Korea between February 20 (when it was described as a broom star about 8 or 9 degrees in length) and March 19. By February 22, the Koreans were seeing a tail of around 15 degrees and "rays radiating in all directions," possibly indicating a degree of structure within the spreading fan not unlike that later described by Angelus.

The Japanese also followed the comet from February 20 until it passed from view sometime after March 20, but few details are provided.

In Europe, the comet is noted as having become visible in the daytime on March 21 and the Italian *Annales Forolivienes* explicitly states that it was visible in full daylight "for about eight days toward the end of March." This text also records the comet as having appeared "in the eastern part of Aries" from late February through early March and to have been visible "for two and a half hours with long hair spreading out," a description consistent with the triangular or fan-shaped tail noted by Angelus.

Mention of the comet is made in English, Austrian, Russian, and Islamic texts, but in general, little extra information about its appearance is added by these chronicles. One English record adds, however, that the tail of the comet "bowed into the north," strongly suggesting a curved Type II dust tail, and an Egyptian text notes that in February, the comet appeared "as large as the Pleiades ... with locks of hair ... [and was] extremely brilliant with light."

It is possible that an oblique mention of this comet is made in Shakespeare's *Henry IV (Part I)*. In the first scene of Act III, the Welsh prince Owain Glyndyr boasts to Henry Percy Hotspur that at the time of his birth "The heavens were on fire, Earth did tremble," to which Hotspur makes the delightful rejoinder "O then Earth shook to see the heavens on fire and not in fear of your nativity." Unwilling to concede the last word to Hotspur, Glyndwr insists that he be given leave "To tell you once again that at my birth the front of heaven was full of fiery shapes."

The association of these "fiery shapes" with the comet of 1402 is apparently based upon a statement of Silas Evans in *Seryddiaeth a Seryddwyr* that the comet of 1402 was described in a poem by Iolo Goch and interpreted as an omen foretelling the success of Prince Glyndwr in his war to restore Welsh independence. Indeed, in June of 1402, the Welsh prince won a significant victory against the English by capturing Edmund Mortimer, the Earl of March. By the end of 1403 he controlled most of Wales and was crowned king of Wales the following year. For a few short years, Wales gained its cherished independence and Glyndwr remains a symbol of Welsh nationalism to this very day.

Whether Goch did mention the Great Comet of 1402 is, however, debatable. According to the *Dictionary of Welsh Biography*, his last ode to Glyndwr was written not much later than 1386!

Actually, Glyndwr was born around 1349; a year with no conspicuous comet. Nevertheless, because of the fortuitous appearance of the comet just before the war turned conspicuously in favor of the Welsh, it is hardly surprising that it should have been seen as an omen for Glyndwr and, as we have already noted in the previous chapter, Shakespeare was not adverse to a little poetic license here and there when it came to dates. The shifting of the comet from the time of an auspicious event in Glyndwr's life to the time of his birth is permissible in plays, if not in historical records!

Before leaving the comet of 1402, we should mention that a record exists of a second brilliant comet that year. This record is included in *Historia Byzantina*, a work by the historian Doukas completed around 1462.

Doukas is writing about the beginning of the battle between Bayazid and Temur on July 28, 1402, when he noted that "At this time ... a sign from the heavens appeared in the western regions as a portent of the evils to come. It was a brilliant comet with its tail more than seven feet high erect like a burning flame, thrusting its beam like a spear from West to East. As the sun sank beneath the horizon, the comet diffused its beam and illuminated the farthermost corners of earth. Nor did it allow the other stars to shine or the night to turn black but instead dispersed its light in a wide arc. The flame was most intense at midheaven while its rays were confined only to the horizon itself. Indians, Chaldeans, Egyptians, Phrygians, and Persians saw this sign and so did the inhabitants of Asia Minor, and the Thracians, the Huns, the Dalmatians, the Italians, the Spaniards, and the Germans as well, and any other nation dwelling along the littoral of the ocean." Doukas adds that the comet remained visible "until the autumnal equinox," implying a visibility of about 3 months. It is also noteworthy that he described the comet as "[diffusing] its beam" after "the sun sank beneath the horizon," seemingly indicating that it was first seen in broad daylight.

There are serious problems with this account, however.

For one thing, it is very strange that a comet as widely observed as Doukas states should have so completely escaped the records of the countries he enumerated. Secondly, the description of the comet's brightness is clearly exaggerated. How many comets are so bright that the stars are blotted out by their light?

Thirdly, overlooking the extravagant element of the description, this comet sounds remarkably like the one earlier in the year; a comet about which Doukas is silent, by the way. This surely raises the suspicion that the comet Doukas describes is, in actual fact, the February/March one, probably displaced to add extra significance to the battle, which was, after all, his primarily interest.

The records of C/1402 D1 are not sufficiently detailed to allow a reliable orbit to be computed. Nevertheless, in 1877 Hind calculated a low-precision one that at least accounts for its general path through the skies. According to this orbit, the comet reached perihelion on March 21 at a distance of 0.38 AU from the Sun. Minimum distance from Earth would have been (according to the Hind orbit) on February 20 at a not-very-close 0.71 AU. Clearly, the orbit alone cannot explain the great brilliance of the comet. According to Vsekhsvyatskij, it must have possessed an absolute magnitude close to 0 to account for its reported luster.

Comet C/1471 Y1

Unlike the Great Comet of 1402, whose spectacular appearance was largely due to its unusually high intrinsic brightness, C/1471 Y1 (the Great Comet of 1472) was set apart from the rest primarily because of a very close approach to Earth, simultaneously coupled with a favorable location in the sky. Intrinsically, it was moderately bright by the standards of comets, though much dimmer than its 1402 predecessor.

Widely recorded in the annals of many cultures, most records agree that the comet first became visible during the latter half of December 1471. According to the English *Chronicles of the White Rose of York*, it was first seen "two or three hours before the sun's rising ... four days before Christmas." Located in the constellation of Virgo, in the south-southeast, it must already have been an impressive sight. A contemporary Polish record described it even at that early stage of its apparition as "a very great comet ... in the latter part of Virgo and Libra."

A Russian text (the *Nikonian Chronicle*) referred to the comet as "a great star," behind which there appeared "a very long and wide ray." This text adds that it "shone very much, more than the [other] stars [and] came out at the sixth hour of the night." The tail, according to this report, "was extended like the tail of a great bird."

It is interesting that the Russian text then goes on to note that "In the month of January, after [January 6], there appeared ... another star with a tail above the southwest. Its tail was very thin and not very long, and its shining was less bright than the ray of the first star. The first star also moved toward the west and would appear three hours before the rise of the sun, and the second one would appear in the same place three hours after sunset." This probably referred to the same comet visible before sunrise and, less favorably, again following sunset.

This morning and evening visibility is also implied by the Egyptian text *Bada'I'al-zuhur fi waqa'I' al-duhur*, which states that during the January/February period "there appeared in the heavens a star with an elongated tail and it would appear from the direction of the west; then it would reappear from the direction of the east."

In Japan, the comet was first seen on January 2 in the region of Coma Berenices. Leo, and Virgo, and was described as being about 4–5 degrees long. It later "turned to appear in the west" and remained visible until at least February 10.

In China and Korea, C/1471 Y1 was followed from early January until well into February. The first Korean sighting was made on January 6, and within 4 days the tail was measured as around 15 degrees in length. This length of tail seems to have been maintained until the middle of January, except (curiously) for January 11, when only 4 or 5 degrees were reported. Maybe skies were hazy on that night!

Between January 15 and 20, the tail lengthened to around 30 degrees, according to Korean observers, before contracting back to half that length and remaining at that value until the end of the month. In remarkable agreement with the mid-January measurements, Italian philosopher and physician Angelo Cato de Supina estimated the tail length as 36 degrees on January 14. According to Cato, the tail was then some 4 degrees wide, and the head of the comet "almost as large as the moon." About this time, it was also being observed by German mathematician Johannes Regiomontanus, who estimated the tail as being 20 degrees long on the 20th and (probably on this same date) gave a measurement of 11 min of arc for the "head" (presumably the central condensation) and 34 for the coma.

The Chinese picked up the comet as a broom star on January 15 (rather later than one would have expected) and about that time also agreed that the tail was 30 degrees in length. By January 20, they simply recorded that it "stretched across the heavens."

It is interesting to note that around January 21 or 22 (the actual date is not given, but the context suggests that time of the month), the Chinese mention that the comet "even appeared at midday." According to the orbit computed by Giovanni Celoria in 1921, it should have passed by Earth at a mere 0.07 AU on January 22. At that time, it was moving through the northern circumpolar constellations at a rate of about one degree per hour (as determined by Italian observer Paolo Toscanelli), as it passed within 15 degrees of the north celestial pole. So close to Earth, the comet must have displayed a very large and diffuse coma (cf. Cato's remark that it was almost as large as the Moon as early as January 14 and Regiomontanus' estimates for central condensation and coma). It is likely that only the central condensation would have had enough intensity of light to be visible in the daytime, but the total light of the extended coma would almost certainly have been at least one or two magnitudes brighter than the central condensation alone. Even in a very clear sky, the latter must have been at least -2 or -3 (Jupiter has been seen in daylight at about -2), indicating a total brightness for the entire coma in the $-4/-5$ range! What a magnificent spectacle the comet would have been, visible all night close to the celestial pole!

After its close encounter with Earth, the comet faded, and the apparent length of its tail shrank. According to Korean observers, the tail declined from about 13 degrees on February 1–9 or less on the 6th, 7–8 degrees on the 9th, and just 6 degrees two nights later. They ceased following the comet on February 21, 4 days after the final Chinese observation had been recorded. In Europe, Toscanalli had already lost it after January 26, but Regiomontanus continued to follow it "until the last days of February," after which it disappeared into the twilight "with the stars of Cetus."

According to the Celoria orbit, the comet passed perihelion at 0.49 AU of the Sun on March 1. As it moved away almost in the direction of the Sun, and beyond it, no further observations were possible.

Regiomontanus attempted to find the distance of this comet by using the method of parallax. In agreement with the prevailing Aristotelian theory of comets as atmospheric phenomena, he estimated its distance to be at least 8,200 miles (13,120 km) and, from this, estimated the central condensation as 26, and the entire coma as 81, miles (41.6 and 129.6 km, respectively) in diameter. These values, of course, fail by orders of magnitude, but he is to be commended for this attempt at determining the physical dimensions of a comet.

Comet C/1577 V1

We now come to one of the more historically significant and better known comets to appear at the dawn of the modern scientific age.

Having passed, apparently unseen, through perihelion (0.18 AU) on October 27, this comet was allegedly seen from Peru on November 1 and from Mexico on the 4th and 6th of that month. These reports are, however, vague and known only from secondary or tertiary sources. Nevertheless, if they are reliable, they testify to the great early brilliance of the comet. The Peruvian report in particular alleges that it shone through clouds like the Moon!

The first definitive records were those of the Japanese, who initially noticed it in the evening twilight of November 8, when the tail may have been as long as 75 degrees, and the comet judged "as bright as the Moon." This brightness is hardly to be taken too literally, but the point has nevertheless been made. The comet was truly brilliant!

The text then adds the rather curious observation that, following its discovery, the comet assumed the form of the character "ta," described by Ho Peng Yoke as having the shape of "a man standing with legs opened and arms stretched, both sideways." This, it must be remarked, is a very peculiar shape for a comet! We must wonder if the figure was upside down, at an angle to the horizon, or upright? If the former, it might describe a tail splitting into two components some distance from the head (the "opened legs" of the character), but the "outstretched arms" are difficult to identify.

At the risk of jumping ahead in our story, it may be worth mentioning that Tycho Brahe described the tail of the comet as being "curved over itself somewhat in the middle," and German observer Michael Moestlin noted that the tail was narrower than the head at its beginning but widened out as the distance from the head increased. It is not clear what being "curved over itself somewhat in the middle" actually implies, but it does at least hint at some change in the appearance of the tail, and this *may* have given the appearance of outstretched arms –with a little imagination supplied!

Moreover, by November 28 the tail had become double, according to Cornelius Gemma of Belgium. If the two tails separated out toward the end of the month, the appearance of spreading legs might have been more apparent, although the man would need to have been standing on his head!

If the comet's head was larger than the emerging tail, and then this rather suddenly bulged out, the appearance of outstretched arms might also have been suggested; once again helped by a liberal dose of imagination.

In any case, the tail does appear to have shown some type of structure. It also seems to have been very intense. A graphic description was given by the German Bartholemaeus Scultetus of the comet's appearance on November 10. It appeared, he said, as a "huge shining spherical mass which vomited fire and ended in smoke." Hardly a detached scientific account, but suitably descriptive nevertheless!

Late in the afternoon of November 13, Tycho Brahe was trying to catch some fish for supper in one of the ponds near his observatory on the island of Hveen (Denmark) when his attention was drawn to an object located between the low Sun and the crescent Moon in the western sky. The object appeared as a star-like point, closely resembling the planet Venus when similarly viewed in daylight – except that Tycho knew the planet Venus was nowhere near that location at the time!

Following sunset, the bright point of light became surrounded by a coma, which Brahe judged to be 8 min of arc in diameter, from which emerged a tail of 21 degrees 40 minutes length and 2.5 degrees width, according to Brahe's rather precise estimates. He described the head of the comet as round and yellow (typical for bright comets observed close to the Sun) and the tail as being formed of "red rays." Notice, once again, the implication of structure observed within the tail.

The following evening, Brahe again saw the comet and noted that it had moved 3.5 degrees during the intervening 24 h. During the following nights, he continued to measure its position whenever possible, noting that its movement slowed continuously from 3 degrees during the 14th/15th interval to just 1.5 degrees per day by the month's end.

The day following Brahe's initial sighting, the comet was first noted in Chinese annals. Describing it as a broom star, they followed it until about 1month after discovery, but supplied little by way of physical description.

About the same time, it also attracted the attention of Korean astronomers, who described it as an "evil star" that "looked like a broom star but was not a broom star." The people were said to have referred to it as a banner of Chhih-Yu. Neither the significance of this nor the reason for not classifying it as a broom star, is explained in the text. It is not clear just how long the Koreans followed the comet, but it does not seem likely that they observed it past December.

The Chinese note that they observed the comet for a month, implying final observations sometime during the middle of December. However, Japanese records state that it was followed until "the first month of the following year," which on their calendar implies a date of around February 20.

In this respect, the Japanese records conflict with the observations of Brahe and Moestlin. The latter observer could still detect it on January 8 but failed in his attempt to sight it 6 days later. Brahe managed to detect it on January 13, noting that its daily motion had decreased to just 25 min of arc and that it had grown so faint as to be barely discernible. Apparently thinking it too dim for further observation, he made no further attempt to see it until January 26, when a friend wanted to know where it was located at the time of his last sighting. Pointing out the position in the sky, Brahe was surprised to see the comet still faintly detectible! Commenting on this final observation, Brahe wrote that it "had grown so small that one could hardly see it, and for aught I know, it faded soon after, and was gone. ... When I saw it for the last time, it had become almost unrecognizable." These comments, together with the negative report of Moestlin in mid-January and the cessation of Chinese (and probably Korean) observations in December, cast doubt upon the accuracy of the Japanese claim to have followed it until late February.

According to Vsekhsvyatskij, this comet was intrinsically one of the brightest ever recorded, indeed, the brightest of any comet in his catalog having a perihelion distance within Earth's orbit. He derived an absolute magnitude, assuming $n = 4$, of –1.8. This figure does, however, seem somewhat high. If true, the comet should have been magnitude 3.5 when Brahe last saw it. Speaking from the experience of observing large comets with developed dust tails, we must doubt that Tycho would have found an object of this brightness as difficult to see as he states. A comet of third or fourth magnitude can actually be quite conspicuous in a dark sky if it sports a significant dust tail. In fact, the total light from the tail can be greater than the magnitude of the head itself, and it is not unheard of for the tail of a comet to be visible with the unaided eye after the comet itself has faded from view. This author's final naked-eye observations of Ikeya-Seki in 1965 and 2006 P1 McNaught were both dust tail sightings, at a time when the heads of these comets were below the naked-eye threshold. In all likelihood, C/1577 V1 would still have displayed a strong tail in late January 1578 and, if as bright as 3.5, should have been a very easy object to locate. For a keen eyed observer like Tycho Brahe, in an age before mercury vapor street lamps, a comet located 45 degrees from the Sun (as V1 was on the evening of January 26) should have presented few difficulties, even if as faint as magnitude 5.5. For him to say that it was then "almost unrecognizable" suggests that it was probably close to sixth magnitude.

Furthermore, an absolute magnitude as bright as –1.8 implies a total apparent magnitude near –8, not even considering the enhancing effects of forward scattering on November 2, when the comet was already 11 degrees from the Sun. It is hard to believe that this would not have led to widespread daylight observations at the time, yet there are no records of this having happened. Even assuming an intrinsic brightness down by two magnitudes should still have made the comet an easily viewable daytime object throughout the first half of November, so the puzzle as to why Brahe's daylight sighting remains alone is not completely solved. If an absolute magnitude of 0 (still incredibly bright by any standard) is assumed, the comet would have been about –3 at the time of Brahe's daylight sighting, quite possible in view of the low altitude of the Sun at the time.

Historically, the Great Comet of 1577 holds an important place. By making accurate positional observations of this comet, Tycho demonstrated that it had a negligible parallax and (contrary to Aristotle and the generally accepted belief at the time) must be located beyond the Moon. In fact, he proposed that the comet was located at least three times further than the Moon and that it moved in a circular path between the orbits of the Moon and Venus. According to his cosmological model, Earth lay at the center of the universe with the Moon orbiting nearby and the Sun orbiting at a far greater distance. The planets Mercury and Venus orbited the Sun (which was itself orbiting Earth) with the comet held in a circular Sun-centered orbit beyond that of Venus. Unlike these two planets, however, the comet was not a permanent object.

This model did not overthrow the belief in crystalline celestial spheres, as was assumed by Brahe's student Kepler and repeated many times since then. The location of the comet within the great spaces between the Moon and Venus would not necessarily have brought it into collision with these supposed spheres. What it did was effectively overturn the long-held belief that the celestial realm beyond the Moon – the superlunar realm – was a region of unchanging perfection. Actually, the comet's location beyond the Moon would not have come as a complete surprise to Brahe, as he had already established a superlunar location for the supernova of 1572. Both the supernova and the comet just 5 years later clearly showed that (contrary to Aristotle and his followers) changes *do* occur in the celestial realms.

Incidentally, the popular perception today is that the ancient geocentric Aristotelian/Ptolemaic model of the universe somehow exalted Earth to a special place in the cosmos and that Copernicus demoted it. Certainly, Earth occupied a geometrically central role in the earlier cosmology; however, this position was more in the nature of a central sump than a central cosmic throne. The earthly (sublunar) realm was believed to be where the gross matter of the universe settled. Earth was the repository of all that was gross and unstable, changing, and impermanent. By contrast, the superlunar regions were celestial, unchanging, and perfect. What Copernicus actually accomplished was the elevation of Earth, not its demotion. As a planet orbiting the Sun, Earth was now a genuine part of the celestial realm. The logic of this also implied that since Earth was not perfect and unchanging (even though a true member of the celestial realm), change and "imperfection" should be expected in other parts of this region as well. Tycho Brahe's observations of the supernova of 1572 and the comet of 1577, when seen in this context, logically followed from the Copernican model, even though Brahe himself never completely surrendered the geocentric cosmology.

As for the crystalline spheres, although compatible with the supposed orbit of the 1577 comet, they were eventually abandoned by Tycho as being incompatible with the motions of Mars. Figuring that Mars at its closest approach would collide with the supposed spheres, he finally took this radical step prior to 1584, by which time he accepted the possibility of intersecting orbits in a space devoid of crystalline spheres and thereby set in motion the search for a new physics capable of explaining the motions of the planets in a space that was truly empty. Eventually this would lead, through Kepler's discovery of elliptical orbits, to the epoch-making synthesis of Newton's theory of universal gravitation.

But *that* is another story!

It is just possible that the comet of 1577 may explain two Oriental records of what appear to have been magnificent comets, though found only in single records. According to a Chinese text, a comet appeared in the west in the year 1581 during September, lasting 30 days and shining with such brilliance as to illuminate the ground! Curious that no other record of this comet exists.

Then, the Korean text *Chungbo Munhon Pigo* records an almost identical comet for October 1587. This one was also said to have had a "bent" tail and, like the earlier one, to have illuminated the ground. Once again, no other record exists.

It is probably significant that both these texts date from the eighteenth century (1726 and 1770, respectively) and were therefore written long after the events they purport to describe. It may or may not be significant that the alleged comet of 1587 was said to have had a bent tail, somewhat reminiscent of the oddly shaped tail of C/1577 V1. Is it possible that these "comets" were actually grossly misplaced accounts of the 1577 object? Speculative, of course, but food for thought nevertheless.

Before leaving our account of the Great Comet of 1577, a curious episode should be mentioned that, if nothing else, highlights some of the difficulties faced by anyone cataloging ancient comets. Apparently, a claim was made by a certain Leonard Thurneysser of Berlin that he had seen the comet as early as October 19 and followed it for a total of 11 days. If this claim were true, Thurneysser's observations would represent a unique series of pre-perihelic sightings of this object. Thurneysser stated that the comet was first seen in Capricornus, moving to Sagittarius by the final observation, at which time its tail was said to point in the direction of Aquila.

Unfortunately, these positions are completely incompatible with the alleged object being 1577 V1. On the other hand, the positions given by Thurneysser do seem to describe a realistic cometary orbit, yielding perihelion on November 8 at a distance of 0.88 AU from the Sun. At the time of Thurneysser's observations, the alleged comet would have been passing within 0.2 AU of Earth, actually making its closest approach to our planet (0.12 AU) just after the termination of his observations on November 3.

Alas, there is no evidence that anybody else saw this comet, and strong suspicions have been raised as to the veracity of Thurneysser's account. In the opinion of his contemporary Helisaeus Roeslin, Thurneysser's claim was a total fabrication, given only in an attempt to justify his previous prediction that some "star" would rise on October 6. For this reason, the comet has been removed from the 1997 and subsequent editions of B. G. Marsden's *Catalogue of Cometary Orbits*.

Nevertheless, if Thurneysser was inventing fictitious observations merely to comply with his prediction, why did he claim to have only observed the comet from October 19? Would it not have served his purpose better to claim discovery on October 6?

Moreover, nobody at the time knew anything about the true orbits of comets (cf. Brahe's assigning C/1577 V1 to a circular orbit beyond Venus). It seems a little odd that a sequence of imaginary positions should have fitted together into a very realistic orbit. True, there is the problem of no confirming sightings, but that is not unique in the history of comet observation. Recall two objects, each seen by only one person, that were eventually identified as early apparitions of well-known periodic comets. An intrinsically rather faint comet passing close to Earth may have slipped by the attention of most people in the late sixteenth century.

Nevertheless, as this comet – if it did exist – bears no relationship to the one seen the following month, we must simply leave the issue where it stands and move on to our next entry.

Comet C/ 1582 J1

This comet was also discovered by the famous Tycho Brahe on May 12. It was then close to the horizon in the evening sky just 16 degrees from the Sun, shining at a magnitude of at least −1 and sporting a tail some 13 degrees long. Tycho saw it again the following evening at an altitude of just 7 degrees and again on the 17th and 18th when he remarked that the head appeared "very small" and compared it to a star of magnitude 4. Presumably, this comparison was not one of brightness so much as the impression of an almost point-like object. In contrast to his long and important series of observations of the Great Comet of 1577, these few observations of C/1582 J1 mark Tycho's only involvement with this object.

The Japanese, however, had a much better view. First noticing the comet on the day following Tycho's discovery, they described the broom star as having the appearance of "a white cloud or rainbow" and estimated the tail as100 degrees in length. It must have been an exceedingly spectacular object that evening in the northwestern sky, and it is on the authority of these early Japanese accounts that the comet is included in our list of the greatest of greats.

In China, the comet was not seen until May 20 (possibly due to cloudy weather) when it was classified as a broom star with a tail "like a chain." The Chinese observers followed it until around June 9, when it would have been about 30 degrees from the Sun.

The comet was also observed in England by Richard Maddox, who saw it on the evening of May 18 in the constellation of Auriga. Interestingly, Maddox claimed that he had first spotted it 8 days earlier (i.e., 2 days before Brahe's discovery, making Maddox the first to see the comet), but that his attempts to further observe it and determine the direction of its motion on subsequent nights had been thwarted, probably by cloudy weather.

In 1783, Pingre computed an orbit for this comet, giving a perihelion distance of just 0.04 AU on May 7; however, this orbit was calculated on the assumption of identity with a purported comet found on March 10, 1582, by Antonio Santucci in Rome. This supposed identity is highly unlikely, and by omitting the March 10 position, Pingre computed an alternative orbit that turned out to be quite similar to those published during the following century by d'Arrest (in 1854) and Marth (1878). The latter yields a perihelion distance of 0.17 on May 6, and bears some resemblance to that of C/1853 R1 (Bruhns). The orbit of this latter object is, however, slightly hyperbolic, ruling out any possibility of it being a return of the earlier comet. Any true association is, putting it mildly, improbable, and the similarity of orbit nothing more than pure coincidence.

Comet C/1618 W1

The year 1618 saw the arrival of three bright comets, C/1618 Q1 in August and September, C/1618 V1 in November and early December, and C/1618 W1, visible from late November until the latter part of January 1619. The first two were not especially noteworthy, although C/1618 Q1 holds the honor of being the first comet to be observed through a telescope (By Johannes Kepler on September 6). The third comet – the one of greatest interest to us – was a different matter, however, and was widely seen as a brilliant and unusually spectacular object.

Having passed through perihelion (0.39 AU) on November 8, C/1618 W1 seems to have been seen first, deep in twilight on November 23 or 24, by Garcia de Silva y Figueroa, ambassador to the Spanish king in Persia. He described it as being similar to Venus in color and of similar, or slightly greater, brightness. The head was, however, somewhat diffuse.

The Chinese began observing the comet on November 26, but their estimates of the length of tail varied greatly. Some records suggest a tail of only 7–8 degrees, while others imply lengths of up to 75 degrees! There may have been large variations in local conditions at the various observing sites, or maybe some of the imperial astronomers had less than perfect eyesight! Be that as it may, the Chinese astronomers managed to track the comet until January 4 the following year.

In Korea, observations commenced on November 30 and indicate a rather long and narrow tail. A further observation on December 7 speaks about a "blue-white vapor" that stretched from Ursa Major through southern Coma Berenices. Although we cannot be certain, this was probably another reference to the tail of this comet.

Kepler also began his rather long series of observations about this time. He first spied the comet through a break in clouds on the morning of November 29 and followed it on every available occasion until as late as January 7, 1619.

In the Philippines, where it had already been spotted as early as November 24, it was described as a "tailed comet" with a "burning star" marking its head. It was said to have remained visible for 3 months.

In Germany, the comet was followed from December 1 until January 22 by the Swiss Jesuit, Johannes Baptist Cysat. When first seen, Cysat described the head as being greater (brighter?) than stars of third magnitude (seemingly an understatement) and noted that by December 8, the tail had attained a length of 55 degrees. The following day, he estimated the tail's length as a full 70 degrees and on the 16th noted that the size of the comet remained little changed, although it had diminished noticeably in brightness.

Even greater tail lengths were reported around that time. According to John Gadbury, various observers gave tail estimates ranging from 21 degrees through the 50–60 range and even up to 90 degrees. One observer, Christianus Longomontanus of Denmark, traced as much as 104 degrees of tail on December 10. The comet itself, during the time of clearest visibility, was estimated by Gadbury to have been "larger" than Spica though "smaller" than Jupiter.

From December 8, Cysat noted that several condensations were visible within the coma. Whether these were genuine sub nuclei or jet structures observed close to the limit of visibility is not clear, but this observation stands as one of the earliest hints of multiple structures within the head of a comet.

As well as observing with the naked eye, Cysat also employed a telescope. In fact, his final observation on January 22 (the last observation specifically recorded anywhere) was telescopic. From his series of observations, Cysat concluded that the comet was either moving through space in a straight line, or traveling around the Sun in a circular orbit between Venus and Mars.

Other noteworthy observations of this comet were made by English astronomer John Bainbridge and Italian Horatio Grassi. The former followed it from

November 28 until December 26, and his drawings of the tail's extent on several dates during that period indicate lengths of 45 degrees on November 28, 60 degrees on December 7 and 10, and 20 degrees on December 23. Like Cysat, Bainbridge also used a telescope during the second week of December in order to make accurate measurements of the comet's position with respect to two specific stars, both when it was near the horizon and when it was near the zenith. From these measurements, he determined that it was at least ten times more remote than the Moon.

Comparing this comet with the earlier two of the same year, Grassi remarked that it "surpassed the others in magnitude of light and daily continuance, so it was outstanding as long as it remained by reason of its course and life, and it drew to itself the eyes of all." As determined by this observer, the tail reached 60 degrees on December 12.

In agreement with Cysat and Bainbridge, Grassi also concluded that the comet lay beyond the orbit of the Moon. Comparing his observations of its positions with those made at Antwerp during early December, he concluded that any parallax was too small to be measured and that the comet must therefore be located well beyond the Moon.

With the comets of 1618, comet observing entered the telescopic era, although telescopic observations remained secondary to naked-eye ones for a long time thereafter. As we shall see below, over 60 years were to pass before the first telescopic discovery of a comet was made, and it would not be until June 21, 1717, that a comet would remain accessible to telescopes alone. On that night, a small comet was found telescopically by Edmund Halley during an observation of the planet Mars. Unfortunately, this little-known "other" Halley's Comet was not seen again, and no orbit could be computed. Probably the first comet followed throughout its path by telescope alone was the intrinsically faint C/1763 S1 (Messier).

Comet C/1680 V1

On the morning of November 14, 1680, Gottfried Kirch was observing the 23-day-old Moon and the planet Mars from Coburg, Germany, when he noticed a star close to the Moon that had not been listed in the star catalog of Tycho Brahe. While attempting to determine the position of this uncharted star, he happened upon an object that he later described as "a sort of nebulous spot, of an uncommon appearance," which he thought might be either "a nebulous star, resembling that in the girdle of Andromeda" or a comet. As it turned out, the "nebulous spot" was indeed a new comet, and Kirch's serendipitous observation goes down in history as the first comet discovery made with a telescope!

At the time, no tail was observable, and Kirch does not seem to have detected the comet with the unaided eye after locating it in his telescope. His comparison with the Andromeda Nebula does not imply equal brightness, only a very general similarity of appearance. Two mornings later, Kirch again located the comet, which had shifted position during the intervening period and displayed a weak tail visible in his telescope for half a degree.

The comet was brightening rapidly, and by November 20 was observed from the Philippines with the naked eye. On the 21st, it was seen in England and the following morning picked up by the Chinese, who described it as a "white broom star" with a tail of about 1.5 degrees long.

By the end of November, it had evolved into a striking naked-eye spectacle. According to J. D. Ponthio, some 15 degrees of tail were visible from Rome on November 27 and 2 days later, lengths of between 15 and 20 degrees were estimated by Arthur Storer, observing from Maryland in North America. Others noted even greater tail lengths at that time – 30 degrees on the 28th according to an unidentified English observer and 36 on November 29 as seen by an unnamed person in Dresden. By then, the comet was said to have been "larger" than first-magnitude stars, a far cry from the little fuzzy patch that Kirch spied through his telescope just 15 days earlier!

Although observations continued as it plunged into morning twilight during the first days of December, the comet was lost to most observers after December 7 or thereabouts. On December 18, it passed just 0.006 AU from the Sun. With the exception of the very faint minicomet C/2001 N1 observed only in data from the SOHO space-based solar observatory, C/1680 V1 is the only known comet not belonging to the Kreutz Group that can be classified as a genuine sungrazer. We will come back to this later.

Close to the time of perihelion passage, C/1680 V1 became exceptionally brilliant, and there are two records of it having been seen in broad daylight. In the Philippines, it was seen at noonday on December 18, when less than 2 degrees from the Sun. Then, on December 19, according to early Dutch settlers at Esopus, New York, "there appeared an extraordinary comet, which caused very great consternation throughout the province." It was said to have appeared "about two o'clock in the afternoon, fair sunshine weather, a little above the Sun, which takes its course more northerly, and was seen the Sunday night, right after about twilight, with a very fiery tail or streamer in the west, to the great astonishment of all spectators, and is now seen every night with clear weather."

The comet started becoming generally visible again after December 20, emerging in the western evening sky as a truly magnificent spectacle. According to some, the tail possessed a distinct golden hue, adding further to the comet's beauty.

From Sussex, the long narrow tail was said to have extended nearly to the zenith. On December 21, John Flamsteed described it as a beam of light about the width of the Moon and extending straight up from the horizon. Others described it as being wider than a rainbow and very long; up to 70 degrees on the evening of December 22 according to Ponthio, who saw the comet from Italy.

This observer also estimated the tail as 3 degrees wide at its extremity on that same evening.

About the same time, an observer in Utrecht judged the comet's head as equaling a first-magnitude star in brightness. From this head the immense tail streamed some 68–70 degrees across the evening sky like a great ribbon. Such was the extent of the tail that it remained visible in the western sky for 5 h after the head of the comet had set!

According to Robert Hooke in England, the tail of the comet reached 90 degrees, with a width of about 2 degrees, on December 28, while on the other side of the channel Jean de Fontaney noted that through a telescope the comet's head differed "both from stars and planets, being dusky light like a cloud, about the size of the moon, and brighter in the middle than the extremes." He continued that "[there is] no reason to conclude it a planet, because sometimes no nucleus, sometimes many are seen, which sometimes divide, sometimes unite." Most of this sounds like marginal observations of jet activity within the central coma rather than a true breakup of the nucleus, although a degree of fragmentation is possible considering the close approach to the Sun just 10 days earlier.

A more conspicuous jet erupted from the nucleus toward the end of December. This feature was observed by Hooke, who described it as a stream of light issuing from the nucleus "in the manner of a sudden spouting of water out of an engine." This is probably the earliest description of a jet or fountain of gas and dust erupting from an active region on the nucleus of a comet. Such features have been seen during several of the returns of Halley's Comet, including those of 1682, 1832, and 1986. Jets of one form or other are quite frequently seen in bright comets, but ones as conspicuous as those of Halley and the 1680 comet are relatively rare, making this early observation by Hooke a real treasure!

On the last evening of December, Flamsteed remarked that the star-like condensation apparent earlier had been replaced by a "hazy light" that to the unaided eye looked larger than a third-magnitude star. When observed through a telescope, this was found to be surrounded by a "bright haze," presumably the outer regions of the coma.

By January 1681, the comet was showing signs of fading, although the tail remained very long and prominent. According to Flamsteed, the head had become fainter than magnitude 3 by January 5, but he still managed to trace the tail for 55 degrees three nights later. On that same date, up to 75 degrees of tail were recorded by Casimiro Diaz at Manila in the Philippines. Flamsteed estimated the brightness of the head as fourth magnitude on January 13, dropping by a further magnitude six evenings later. At the same time, the tail had shortened to 40 degrees on the 15th according to Isaac Newton, who also noted that it had become curved, "with the convex side . . . to the south."

An interesting, but largely overlooked, observation was made by Kirch on January 7, and repeated by the same observer the following night. Kirch noted that in addition to the "large, bright tail extending away from the Sun," the comet on both nights also possessed a very weak second tail that pointed directly *toward* the Sun. It is unfortunate that a better description was not given of this anti-tail, and that no confirming observations by other people have been found.

During February, the comet progressively dropped from the view of astronomers as it continued to fade. Flamsteed estimated the head as just magnitude 7 on the 4th, although the tail remained visible with the naked eye until well into the month. Newton estimated tail lengths of 6–7 degrees for most of the first week of February, although he traced it out to 12 degrees under exceptional skies on the 6th. A similar tail length was given by Ponthio as late as February 17, although 2 days prior to this estimate, Flamsteed remarked that the comet had become nothing more than "a dusky light" to the naked eye. On the nights of February 19 and 20, Newton could trace just 2 degrees of tail with the aid of a telescope.

At the time of its appearance, there was a widespread belief that the comet of late December and January was not the same as the object that Kirch and others had observed before dawn in November. Thanks mainly to the work of Isaac Newton as he developed his famous theory of universal gravity, the "two" comets were shown to be a single object orbiting the Sun along a nearly parabolic trajectory. The comet actually played an important role in Newton's arguments in the *Principia*, assuring this object a special place in the history of physical science. Newton's special interest in this comet led him to follow it telescopically long after other observers had given up. In fact, his last measurement was not until March 19, by which time it had become "scarcely discernible" in his telescope. Judging by the earlier estimates of its brightness, it may then have been no brighter than ninth magnitude; the faintest at which a comet had been observed until that time and close to 40 *million* times fainter than its observed peak brightness in broad daylight on December 18 and 19!

Earlier, we referred to the fact that this comet is, apart from the one minor exception already mentioned, the only "sungrazer" that seems unrelated to the Kreutz Group of comets. This statement may need modifying. True, the comet's orbital elements show no similarity with those of the Kreutz family (except, of course, for the very small perihelion distance), but that may not be the end of the matter. H. Kreutz himself was the first to draw attention to a very odd coincidence (if "coincidence" is what it really is!) in the orbits of this comet and that of the great Kreutz sungrazer of 1882. The orbits almost intersect! The 1680 comet, the Kreutz sungrazers (all fragments of a single progenitor comet, according to the accepted theory), and the tiny C/2001 N1 are the only objects known to have ventured to within 0.01 AU of the Sun. Surely, the chances that two of these apparently unrelated objects should move in such a way that they can (theoretically, at least) almost collide are just too miniscule and hints at a disruption some time in the past of a far more violent nature than the gentle nucleus splitting that gave rise to the Kreutz Group itself.

Comet C/1743 X1

This comet, regarded widely as the most spectacular of the eighteenth century, was long known unofficially as de Cheseaux's Comet, even though Dirk Klinkenberg of Haarlem actually found it 4 days prior to de Cheseaux. The reason de Cheseaux's name became more widely associated with it is largely because his series of observations was more extensive and enabled him to compute its orbit, plus the fact that his final observation was an interesting and controversial one, about which more will be said in a little while. Today, the more distinctive name of "Klinkenberg-de Cheseaux" is increasingly being given to the comet, although only the designation is truly "official."

Klinkenberg found the comet of December 9, 1743 and Jean Philippe Loys de Cheseaux (Lausanne) on December 13. The latter described it as equal in brightness to a star of third magnitude and already exhibiting a tail. Through a telescope, de Cheseaux compared it to a small nebulous star about 5 min of arc in diameter. However, despite (or maybe because of!) its relatively small size, he remarked that it was easier to see than the Andromeda Galaxy.

The comet's brightness had increased to second magnitude by December 21, as estimated by Jacques Cassini, who observed it from Paris. According to this observer, when seen through a telescope, it then resembled a nebulous star larger than Jupiter.

In days when information was disseminated a lot slower than today, news of comet discoveries spread only gradually between countries, and as C/1743 X1 became more conspicuous in early 1744, several new "discoveries" were made. For instance, an observer in Cambridge, England, found it on January 3 as a new object. Another "discovery" was made on the night of January 8.

In China, the comet was first seen on January 4 and described as a broom star, yellowish-white in color and "as large as a pellet." A tail measuring 1.5 degrees pointed in an easterly direction. The Chinese followed the comet until February 25, by which time it had covered "a distance of more than 29 degrees."

The comet entered Pegasus on January 14 as it drifted slowly westward through the evening sky. The rate of brightening appears to have been quite slow at that time, and most observers throughout the month described it as resembling a second-magnitude star. The tail seems not to have been especially intense, although by January 17 it had reached "about 6 or 8 degrees long" in the estimation of English observer Thomas Sparrow.

An interesting observation of what appears to have been a sunward jet or fountain of material erupting from the nucleus was reported by G. Heinsius (St. Petersburg) on the night of January 25. He described the feature as a "triangular shining ray" having the comet's nucleus as its tip and extending its base toward the Sun. The appearance of the feature was such that "The lateral borders of this needle seemed curved as if they were pushed from the inside to the outside by the action of the Sun."

Fig. 4.1 Comet Klinkerberg–de Cheseaux over St. Martin's in the Fields Church, London, January 26, 1744 (courtesy, Guildhall Library and Guildhall Art Gallery, City of London)

The comet's development became more significant during the following month. By February 7, the tail had reached a length of 20 degrees, according to Cassini. Observing through a telescope, Cassini described the comet's head as being "nearly round" on February 9 but elongated in the direction of the tail two evenings later. On the 13th, G. Smith remarked that the comet had become brighter than any star in the night sky except Sirius and that the tail then exceeded 25 degrees in length. The tail had become double, according to Cassini on the 15th, with the "western branch" reaching 24 degrees and the "eastern branch" 7–8 degrees long. This observation may indicate the existence of both plasma and dust tails, as is common with large and bright comets. A Type II dust tail is strongly implied by Cassini's description on February 23 of a curved tail with the convex side facing west. By then the comet's brightness exceeded that of Jupiter and was starting to rival Venus! Smith saw the comet together with Venus in the morning sky on February 25 and remarked that the two objects were comparable in brightness and that both remained visible after all the stars had been lost in twilight. He also commented that the tail formed "a sort of curve to the west."

On that and the following day, Italian observers Gianpaolo Guglienzi and Gianfrancesco Seguier saw the comet in the evening, very briefly near the setting Sun, as well as just minutes before sunrise on February 28. At midday on that same date, these observers located it both telescopically and by naked eye in broad daylight. This was the first time that a comet had been observed telescopically in the daytime. It was also seen on that day by James Bradley (Oxford), who observed it only a short time before the Sun's transit.

On the morning of March 1, while the comet's head was too deep in twilight to be seen, the tail extended from beneath the horizon to a height of about 15 degrees. That was also the date of the comet's perihelion passage (at 0.22 AU from the Sun), after which it slipped quickly into the southern hemisphere. From a ship off the coast of Western Australia, Pascoe Thomas saw it just before Sunrise on March 3, estimating its tail as 10 degrees long and the head so bright that it remained visible even when the Sun was "about one diameter above the horizon."

From March 5 until March 9, the extremity of the tail was observed by several European observers as a multiple system appearing in the eastern sky before dawn. De Cheseaux's own observations of this phenomenon are the most widely known, and it is probably due as much to these, his final, sightings of the comet as to its discovery, that his name has been so closely associated with it. We will say more about these interesting and important observations shortly.

By March 17, the comet (then an exclusively southern hemisphere spectacle) sported a tail of nearly 40 degrees according to Thomas. Dutch navigators sailing south of Madagascar also saw it on the mornings of March 18 and April 22. On the former date, they apparently traced the tail for some 80–90 degrees. No observations later than April 22 are known.

For many, the most interesting feature of this comet was the so-called multiple tail observed in early March. The best known observations of this feature are the ones made by de Cheseaux himself on the mornings of March 8 and 9. His classic illustration shows six complete tails rising up from beneath a weakly dawn-lit horizon, with dotted lines tracing what he believed to be the path of the tails beneath the horizon as they diverged from the comet's head. The impression given has been likened to a Japanese fan. Each tail is drawn as a separate appendage and they are all given a slightly different curvature as they diverge from a common origin.

Nevertheless, we should remember that this drawing is only an impression of what he *thought* it would have looked like had the whole comet been visible. Neither de Cheseaux nor anyone else actually *saw* the comet looking like that. Nobody reported multiple tails when the entire comet was visible in a dark sky. The nearest to a multiple-tail system was the double tail reported on February 15; not at all unusual for a large comet. It is unfortunate that a number of more recent astronomy books have taken de Cheseaux' drawing and given "artist impressions" of what they supposed the comet looked like against a dark sky. There are a number of these "impressions" of a great six-tailed unfurling fan hanging resplendent in a nighttime sky, and it is easy to simply assume that it really did look like that. But as we read through the actual eyewitness descriptions of this comet, we find nothing that even hints at this popular characterization. What we find is the account of a spectacular comet with a large and strongly curved Type II dust tail. Note, for instance, that Thomas made no mention of multiple tails in his description of the comet on March 3, yet (as we will see below), the first recorded sighting of multiple rays came only 2 days later.

But back to the drawings of March 8 and 9. These were treated with a good deal of skepticism by later generations of astronomers. Not only did they seem so peculiar, but confirmation from other sources initially appeared lacking. As it tuned out, however, other lesser known records did exist, and when these were rediscovered by the wider astronomical community, the de Cheseaux drawings essentially became the defining characteristics of this comet.

The "confirming" observations were those of Margaretha Kirch at Berlin on March 5, Joseph-Nicolas Delisle (St. Petersburg) on the 6th, and those of both these observers together with Gottfried Heinsius the following morning. A copperplate engraving of the phenomenon as seen from Berlin on the latter date (presumably by Kirch) shows 12 distinct features, one of which is itself divided into three individual rays. Apparently de Cheseaux's observations on the following two mornings only caught the dwindling remnant of the spectacle! The features shown on the Berlin engraving, however, do not have the same "tail-like" appearance as those drawn by de Cheseaux.

So what were these peculiar structures? They were almost certainly not tails in the true sense of the word. For one thing, the change in their number from March 7 to March 8 implies something much more ephemeral than dust tails. More importantly, their orientation made no sense. They did not point even approximately away from the Sun. In fact, they were almost perpendicular to an imaginary line connecting the Sun and comet! They would be better explained as features running *across* a single curving dust tail, not individual tails projected away from the Sun.

But what features run more or less across a comet's tail?

The answer ... *striae*!

A broad and strongly curving dust tail, heavily striated, could account for the phenomenon observed by these astronomers. The Great Comet of 1744 was not a six-tailed dragon, but a remarkable example of a heavily striated and strongly curved Type II tail, something that fits far more readily with the other descriptions of this object.

We have already mentioned similar observations reported for the comets of 1264 and 1402, but the one that duplicated the de Cheseaux phenomenon most dramatically appeared as recently as January 2007. The magnificent Comet McNaught has actually had the unexpected consequence of arousing more interest in the 1744 comet than ever before, thanks to some remarkable observations in mid/late January from mid-northern hemisphere latitudes.

At that time, the comet had passed too far south to be seen from the northern hemisphere, yet throughout middle northern latitudes, numerous people both saw and photographed a series of 'fingers of light" – sometimes described as looking like the beams of a stationary aurora – rising out of the darkening evening twilight. The nature of these features was quickly identified. They were striae in the comet's enormous and strongly curved dust tail. From southern latitudes, McNaught appeared in the southwest with a great tail that arched

Fig. 4.2 A repeat performance of the phenomenon observed by de Chesseaux, Kirch, et al. in early March 1744! This photograph shows striae in the tail of the recently set Comet McNaught projecting above the horizon on January 18, 2007 (courtesy, Terry Lovejoy)

Fig. 4.3 A similar view of tail striae of Comet McNaught photographed from near Wellington, Colorado, on January 19, 2007. At that date, the comet itself remained beneath the horizon, very closely imitating the circumstances encountered by de Cheseaux and Kirch in March 1744. © 2007 Mary Laszlo, used with permission

over the point of Sunset before dropping down below the horizon in the northwest. The entire tail could not be seen from either hemisphere; the region below the northwestern horizon was only visible north of the Sun to observers in the northern hemisphere. Being heavily striated right to its extremity, this "northern" section of tail was visible as a series of striae orientated upward from the horizon. It was when the photographs of this phenomenon were compared with Kirch's and de Cheseaux's depictions of the 1744 comet that the striking similarity was noted. It was almost as if these pictures had been taken back in time to 1744!

There can be little doubt that the McNaught and de Cheseaux phenomena were of the same kind. After 263 years, the mystery of the "six-tailed comet" can finally be laid to rest!

Comet C/1769 P1 (Messier)

Following his observations of Halley's return in 1759, Charles Messier of the Marine Observatory in Paris attempted to improve his positional determinations of this comet by making more accurate measurements of some of the reference stars used. Halley's Comet had long gone, but during the course of his reference-star measurements in January 1760, Messier accidentally found a new comet with his telescope. This appears to have set him on a course of deliberate telescopic sweeps for new comets, and, as they say, the rest is history. Messier became the forerunner of the numerous astronomers – amateur and professional – who ever since have spent their nights scanning the skies for these elusive fuzzies. In the years following, Messier became known as the "ferret of comets" and eventually had his name given to 12 of these objects, plus a 13th in 1770 now officially named for the mathematician who determined its unusual orbital history (Lexell).

Of all his comet discoveries, however, none can compare with the magnificent object of 1769.

Messier swept up this comet on the night of August 8, 1769, describing it at the time as a nebulosity several minutes of arc in diameter that "appeared faintly in the telescope." It could not have been *too* faint, however. The following night, when he confirmed by its movement that it was indeed a comet, Messier also spied it with the naked eye. On the 15th he recorded a tail length of 6 degrees with the naked eye and through a telescope measured the coma as 4.5 min of arc in diameter, with a "nucleus" (actually, what we would today call a central condensation) 86 s of arc across.

Because the comet was in the morning sky, moonlight interfered with observation following a full Moon on August 17. Still, naked-eye observations were made by G. D. Maraldi and Cesar Francois Cassini de Thury at Paris on the 22nd, and on the 24th the comet was first recorded as a broom star by the astronomers of China.

The following morning, it was also observed by Jean de Surville from on board the *St. Jean-Baptiste*, sailing near the Philippines. He described it as being "bearded . . . not bright."

As the Moon waned in the morning sky, the comet's tail grew more obvious. On the morning of August 27, the Chinese described it as being "as large as a pellet" with a tail about 4.5 degrees long. In a moonless sky the following day, Messier traced the tail for a full 15 degrees, noting that the comet and its nucleus appeared more brilliant than they had on previous mornings.

On board the *Endeavour* in the South Pacific, Captain James Cook recorded for August 30 that he had observed a comet "a little above the horizon in the eastern part of the heavens" and judged its tail as 42 degrees long. The next morning, Maraldi and Cassini noticed that the tail appeared slightly curved. Curiously, these astronomers traced the tail for just 18 degrees. Clearly, local atmospheric conditions were critical to how much of the tail could be seen.

The comet was widely observed from ships at sea during late August, but little extra information is supplied by these accounts.

According to Messier's observations, the tail length was 36 degrees on September 3, 40 degrees the following day, and over 43 degrees on September 5. At that time, he remarked that it appeared curved, with the convex side facing north, and had become "very feeble" at its extremity. Messier also noted that the nucleus had begun to show a hint of red or orange color.

The following day, tail lengths of 70 and 49 degrees were reported by E. Zanotti, Maraldi and Cassini, and Messier. Messier added the important information that the tail was not homogeneous, but instead composed of bright luminous rays running parallel to one another. This sounds like the visual description of rays in a bright Type I plasma tail.

The tail continued to lengthen, and on September 9 was measured as 55 degrees by Messier.

The comet made its closest approach to Earth on September 10 at 0.32 AU. Thanks to very favorable observing geometry, the visual tail reached prodigious lengths around that time –60 degrees as determined by Messier on that day and an amazing 90–98 degrees as seen by Pingre and de la Nux from a location at sea between the Canary Islands and Cadiz. Pingre added that the extremity of the tail was so faint that several degrees were lost to view when Venus rose, but whether this was due to the light of that planet or the first hint of approaching dawn is not clear. Either way, the more distant reaches of the tail were obviously very faint.

The tail remained very long for several days after the comet's closest approach. According to Zanotti, it was 74 degrees on the 13th, while Stiles estimated 90 degrees the following morning. Although its more distant parts were faint, observers noted that the first 40 degrees were very bright.

As the comet headed into twilight and toward perihelion (0.12 AU on October 8), it became increasingly difficult to see, and the visible tail rapidly contracted. On September 15, 40 degrees were still being reported, but just 2 days later Messier managed to trace it for a mere 2 degrees in bright twilight.

The comet itself remained easily visible, however. According to Messier, the "nucleus" at that time was surrounded by a kind of "atmosphere" extending back into the greatly diminished tail. Searches by Messier on the 17th and 18th failed to find the comet, although Maraldi apparently glimpsed it "within the rays of the Sun" on the latter date. Presumably this implied a sighting in very bright twilight rather than daylight per se, but the sky was too bright for him to accurately determine the comet's position.

After perihelion, the comet remained hidden in twilight until its recovery on October 23 by astronomers at the Royal Greenwich Observatory in England. Having moved well away from Earth as well as receding from the Sun, the comet was, by then, but a shade of its former self. Nevertheless, it was visible with the naked eye – albeit with difficulty – when Messier again spotted it the following day low in the evening twilight. As seen through a telescope, the nucleus appeared very bright, although only 2 degrees of tail were visible.

The comet was better placed on October 26, when Messier estimated its nucleus as equal in brightness to a third-magnitude star. Both Messier and Maraldi continued their observations during late October, describing the tail as becoming "long" as the comet emerged higher in the evening sky, but they gave no actual measurements of this length. By contrast, Chinese astronomers, who also recovered the comet at the end of October, described it as being "small," with tail lengths of just 1.5 degrees.

On the evening of November 1, Messier seems to have had a particularly good view of the comet and estimated the tail at that time as 6 degrees long and "very noticeable." Three days later, the comet was about third magnitude, still visible to the naked eye and sporting a "very apparent" tail brighter down the middle than at the edges. In moonlight on November 8, the tail had shrunk to only 2.5 degrees.

After the Moon left the evening skies, Messier again found the comet on November 15 and 17, but only with a telescope. He did manage to glimpse it with the naked eye one last time on the 18th but needed a telescope on the 20th when, nevertheless, he noted the tail as still extending for 2 degrees. He continued to see the comet, telescopically and "with much difficulty" on the 30th when, however, some 1.5 degrees of tail could still be traced.

The persistence of the tail even as the comet faded to obscurity suggests that this was probably a dust tail rather than the predominantly plasma tail described prior to perihelion. The development of a dust tail would be expected for a bright comet that approached the Sun as closely as this one. Indeed, this was exactly how the very similar comet C/1996 B2 (Hyakutake) behaved, as we shall see in Chapter 5.

The final observation of Messier's comet was by P. W. Wargentin on December 3, after which a period of cloudy weather thwarted the principal observers for several days. When the skies again cleared, further attempts to relocate the comet ended in failure. It had finally faded from view.

With the departure of this comet, the precession of "greatest of greats" from the turn of the second millennium until 1800 comes to a close. There were, of course, bright comets later in the eighteenth century, but none that could justifiably be included in our list. What followed the year 1800, though, was a different matter. Great comets, including a goodly number that certainly deserve to be classed among the very greatest of them all, have paraded through the skies from soon after the turn of the nineteenth century until the present day, to the delight of astronomers throughout the world. Let us take a look now at the best of these celestial delights.

Chapter 5
The Greatest Comets from 1800 to Present Times

Even a quick glance through the lists of great comets shows it to be top heavy! The last couple of centuries seem to have had more than their share of very bright ones. (Please do not read this as a complaint!).

In part, this is probably because the opening up and increased settlement of the southern hemisphere has allowed those comets that have saved up their best performance for southern climes to be observed through their maximum brightness. Also, comets were found that remained exclusively southern objects, and probably would have slipped by unnoticed in earlier centuries.

Moreover, this period also saw the return of a number of Kreutz sungrazing comets. These, as we shall see in Chapter 6, progressively fragment, implying that the group is becoming more and more populated by smaller objects. Nevertheless, even though the recent sungrazers are smaller than their earlier parents, many are still capable of becoming great comets. Because the brightness of a comet depends more on the active surface area of its "nucleus" than diameter per se, it is not impossible for the child to outperform the parent, as some of the more recent sungrazers may actually have done.

The most brilliant, and arguably the most spectacular, comets since 1800 were sungrazers, principally those of 1843, 1882, and 1965. Other great sungrazers were seen in the years 1880 and 1887, helping swell the number of great comets during the past couple of centuries. The first three were surely among the greatest of comets, but we will leave them until Chapter 6, when this interesting group will be looked at it more detail.

Nevertheless, the list of great comets since 1800 would still be impressive even without the sungrazing contribution. Even if we place rather tight constraints upon our definition of a "great comet" and omit the three returns of Halley during this period (despite the 1910 return being exceptionally spectacular for those who happened to be in the right place at the right time), we can still find 29 comets that surely qualify as having been great. This is an average of one every 7 years, although they were not so evenly distributed, of course. They were certainly not the only bright comets during that period, and, in fact, if great comet was defined more liberally, several more could be added to the list!

A moment ago, we remarked that the great comets were not evenly distributed throughout this period. There are actually two quite distinct "clumps" of

D. Seargent, *The Greatest Comets in History*,
DOI 10.1007/978-0-387-09513-4_5, © Springer Science+Business Media, LLC 2009

bright comets; one from 1874 until 1887 and consisting of the great comets of 1874, 1880, 1881, 1882, and 1887, together with the "fairly great" Comet Wells of June 1882, and a second clump from 1957 until 1976, comprising the great Comets Arend–Roland and Mrkos of 1957, Seki–Lines (1962), Ikeya–Seki (1965), Bennett (1970), and West (1976), together with the relatively bright Wilson–Hubbard (1961), White–Ortiz–Bolelli (1970), and Kohoutek (1973). This tendency to bunch is typical of a random distribution and is of no physical consequence in itself, although the above-mentioned "bunches" were accentuated to some degree by the three Kreutz comets of the 1880s and Ikeya–Seki plus White–Ortiz–Bolelli of the 1960s and 1970s.

The nineteenth century has long been regarded as a century abnormally rich in bright comets, but it is not so widely accepted that the twentieth was just as productive. Comets that must surely be thought great in anybody's estimation were seen in 1901, 1910 (two in that year, one being Halley's!), 1927, 1947, 1948, 1957 (another two!), 1962, 1965, 1970, 1976, 1996, and 1997. The average is one in a little less than 7 years. Then, just 7 years into the twenty-first century, there appeared another great comet – in fact, not just a *great* comet but one of the *greatest* of comets! Let us hope that the present century keeps up the standard!

Actually, there has been quite an influx of naked-eye comets between mid-1995 and late 2007. Some 24 comets were observed by the naked eye in that period, a remarkable tally. Admittedly, quite a number of these were just within naked-eye range, but others became impressively bright and three definitely qualify for inclusion within the "greatest of greats" fraternity. We will come to these in due course.

Sifting out the comets that seem best qualified to be listed here was not an easy task. Some readers will no doubt take issue with a few of those chosen, and, more especially perhaps, those that have been omitted.

In any case, after a lot of deliberation, the comets here judged best qualified to go into the list of the greatest comets of history are; 1811, 1843, 1858, 1861, 1882, 1910 (January), 1910 (Halley), 1927, 1965, 1976, 1996, 1997, and 2007. Of these, Halley's in 1910 has already been covered in Chapter 2, and those of 1843, 1882, and 1965 are, as already indicated, special cases reserved for Chapter 6.

Before looking at the others individually, let us say a few words to defend the passing over of some objects that many readers may feel should have been included. The principal ones are the great comets of 1874 (Coggia), 1970 (Bennett), and possibly the Great Southern Comet of 1947.

Without doubt, C/1874 H1 (Coggia) was a beauty; a true great comet. At its brightest, it probably exceeded the first magnitude and displayed a series of envelopes within its coma that astronomers compared with those of Donati's Comet 16 years earlier. Suitably placed observers also noted maximum naked-eye tail lengths reaching 70 degrees as the comet passed near Earth in July.

So why, then, was this comet left out of our list?

Principally, the decision to pass it over was based on its poorly placed aspect at the time of maximum tail length. Had it passed by Earth while still at large elongation, it would certainly have warranted inclusion, together with similar objects (such as C/1996 B2) that became great primarily because of a close passage of our planet. But in Coggia's case, the close approach happened as the head of the comet plunged into twilight. At the time of greatest apparent tail length, the head was invisible in deep twilight. The very long tail was described at the time as rising out of twilight like an auroral beam but seems not to have been bright enough to attract wide attention. Indeed, its comparative faintness is implied by the fact that pairs of observers standing together saw radically different tail lengths. On July 17, for instance, the very experienced observers K. Bruhns and J. Schmidt gave 36 and 54 degrees, respectively, C. Abbe saw 30 degrees, while a friend standing next to him saw 60! Five nights later, E. Trouvelot watched the tail for more than an hour, stating that it would appear to come and go, one moment being clearly visible and the next vanishing altogether, apparently due to slight differences in atmospheric transparency. The experiences of these observers imply that the tail, though very long, was nevertheless so faint that its visibility depended critically on both atmosphere and eyesight. This hardly compares with the spectacles provided by comets such as 1264 or 1996.

The comet had a perihelion distance of 0.68 AU and (according to Vsekhsvyatskij) an absolute magnitude of 5.7. This combination does not really amount to the "right stuff" for an outstandingly great comet. Please note that this is *not* denying that the comet was a great one. It was. It simply was not one of the *greatest* of its class.

Similarly Comet Bennett, though a magnificent object and great by any reasonable definition, was not sufficiently outstanding to be included with the likes of those listed here. It was not the most spectacular comet ever seen, but it was well placed and bright for an extended time and the intense curving dust tail and often highly contorted ion tail, made it both visually and photographically a very fine specimen. It may not have been awarded a gold medal in the Great Comet Olympics, but it certainly deserved a silver!

As for the Southern Comet of 1947, this seems to have been very spectacular for a short time and was even described as looking like the full Moon with a tail by a correspondent to *Sky & Space* magazine recalling the experience of many years earlier! Once again, though, it is doubtful if it should be included here, although its "greatness" is certainly not in dispute.

Moving on from these few words about comets that have not been included, and the reasons for omitting them, let us now look at the ones that were included, beginning with an object that has become almost legendary among comet observers and which until as recently as 1997 held the record for the longest period of naked-eye visibility of all recorded comets. This was the comet considered by Napoleon I of France as a good omen favoring his planned invasion of Russia (he got that wrong!) and even gave its name to a vintage of wine upon which it was supposed to have had a beneficial influence. We refer, of course, to the Great Comet of 1811.

Comet C/1811 F1 (Flaugergues)

On March 25, 1811, H. Flaugergues of Viviers in France discovered a comet in Argo Navis, a huge constellation in the southern regions of the sky, which has since broken into several smaller component stellar groupings. Although re-observed by its discoverer the following night and again on the 28th, 29th, and April 1, no official confirmation of the comet was made by other observers until April 11, when it was sighted by F. X. von Zach, who had formerly received word of its presence, and accidentally by J. L. Pons, who had not. Around the middle of April, naked-eye observation began and continued until the first week of January 1812, something hitherto unprecedented in the annals of comet observation and not exceeded until Comet Hale–Bopp of 1997.

The comet brightened slowly during April, May, and June but from early June became increasingly difficult to see as it drifted into twilight. The final observation prior to conjunction appears to have been by Alexander von Humboldt at Paris on June 16, deep in evening twilight. It remained hidden until August 18, when again located by its discoverer as a difficult object very close to the horizon. On the morning of August 22, it was also spied very near the horizon by Olbers, who noted that the nebulosity had "brightened toward the middle," but that the sky was too bright to tell whether it had developed any semblance of a tail.

On the evening of the same day, J. E. Bode also found the comet with the aid of a telescope just before it went out of sight beneath the horizon. A few hours later, before dawn the following day, he again saw it and noted that it was then visible with the naked eye. For the first time, a tail was detected, although Bode only wrote that it was "short."

The comet must have become reasonably bright, however, as F. W. Bessel saw it with the naked eye when just 4 degrees above the horizon on August 23. The tail also continued to develop during this time, with Olbers tracing it for 3 degrees on the 29th as the comet finally began to clear the evening twilight. Oddly enough, although the coma was becoming quite strongly condensed, several observers remarked on the absence of a true nucleus within the central region.

The comet was clearly visible in both the evening and morning skies of early September, becoming bright enough to draw wide attention. It was noted, for instance, by Alexander Ross while traveling down the Columbia River, OR, in the United States on September 1. Ross wrote that it was visible almost due west, about 20 degrees above the horizon, and "very brilliant ... with a tail about 10 degrees long." He noted that the Native Americans were also aware of the comet's presence, regarding it as a sign that the Good Spirit had sent the explorers. Because of that association, the explorers were received with "a reverential awe." It should be remarked that this happened on the night before a full Moon! Clearly the comet and its tail were then very bright.

Apparently, the nucleus had also become quite prominent by September 6, as reported by Matthieu in Paris, who also mentioned that the tail (5–6 degrees long) had divided into two close branches. Throughout the following days, as the Moon continued to wane, many observers reported tail lengths of around 10 degrees.

Perihelion occurred on September 12 at a distance of 1.04 AU from the Sun. This is a remarkably large perihelion distance for a great comet and testifies to the high-intrinsic brightness of this object ($H_{10} = 0$, according to Vsekhsvyatskij's determination!).

In Glasgow, William Herschel observed the comet's head through a telescope under a magnification of 110 on the night of September 18. He described it as looking very like a globular nebula some 5 or 6 min of arc diameter "of which one or two minutes about the centre were nearly of equal brightness." No doubt this refers to a small disc-like central condensation or false nucleus within the central coma. The tail, as seen by Herschel on that same night, was 11–12 degrees long and "towards the end . . . its curvature had the appearance as if, with respect to the motion of the comet, that part of the tail were left a little behind the head." Through a night glass with a field of 4 degrees 41 min, the tail was seen to be accompanied by a "stream" on each side. These two streams became progressively spread out and diffuse as they moved away from the head, eventually merging into a single featureless sheath of light at the tail's extremity. On the same night as Herschel made these observations from Scotland, the comet was also noted far away in Yucatan, Mexico, apparently for the first time.

On the 6th of the following month, Herschel noted both an inner and an outer coma making up the comet's head. The first was measured as 3 min 45 s, while the latter surrounded it as a fainter envelop some 15 min of arc in diameter. The tail's length was measured as 25 degrees on that same night.

Throughout October, tail lengths of from 12 to 25 degrees were observed by experienced astronomers such as G. Piazza, Olbers, and Herschel. The latter continued to observe the two streams defining the edges of the tail within several degrees of the head. On October 12, he remarked that these remained well defined for about 6 degrees, "after which their scattered light began to be pretty equally spread over the tail," which he measured at that time as 17 degrees long. He also noted that the tail was widest (about $6\frac{3}{4}$ degrees) some 5 or 6 degrees from the head, after which it narrowed somewhat. It is interesting that this widest part coincided with the point at which the two side streamers lost their identity and merged with the general glow of the tail, although Herschel was apparently not impressed by this coincidence.

There are, unfortunately, very few brightness estimates of this comet. The best indication of its luster was probably on October 20, when Schroter saw it at a time when only Vega and a few other very bright stars were visible in twilight. This observation implies a brightness of at least 0 magnitude. At the same time, the head was estimated as between 20 and 28 min of arc in diameter, making it at least as large, in real terms, as the Sun itself!

Closest approach to Earth occurred on October 16 at a relatively distant 1.22 AU, and by late that month, the comet was clearly fading and the tail diminishing in size. On November 4, it was described as "a faint nebulous star of the third magnitude" by W. J. Burchell, who saw it while camping near the Vaal River in South Africa. The next day, from Scotland, Herschel found the tail to be down to 12.5 degrees, with the "preceding stream" 5 degrees and 16 min and the "following [stream]" 4 degrees and 41 min long. On the 9th he made note that the tail looked very similar to the Milky Way "in places where no stars can be seen." It continued to shrink, being down to 7.5 degrees on November 17, 6 degrees 10 min on the 19th, and "hardly 5 degrees and of a very feeble light" on December 2.

Bode also estimated the tail as 5 degrees on December 3, and Herschel implied a similar length on the 9th and again on December 14. From the 9th the two side streamers had essentially lost their identities.

The comet drifted toward evening twilight in early January and according to Ferrer was scarcely discernibly with the naked eye around the end of the first week of that month. Nevertheless, some 3 degrees of tail could still be traced as late as January 8, according to J. H. Fritch in Germany, although it seems that naked-eye sightings ceased soon thereafter.

The comet was lost to view by mid-January and remained unobservable until July 11, when Ferrer recovered it in a 10-cm (4-in.) refracting telescope with a magnification of just five times and a field of 5 degrees diameter. He described it as "a very slight vapor, its tail opposed to the sun scarcely looked 10 minutes in length." It was last seen as a very faint object by V. Wisniewski in Russia on August 17, 1812, almost 17 months after its discovery. At that time, this was the longest that any comet had been kept in view.

Before leaving our account of this comet, we should draw attention to a couple of urban legends that have unfortunately gathered around it, with some observations stated as fact in quite reputable places that almost certainly have no grounding in reality. The first is the supposed daylight visibility of this comet, which we find mentioned from time to time. There surely can be no truth in this. It is *just* possible that it could have been found telescopically during daylight hours, but as far as we are aware there are no records of this, or even of any attempts to find it in the daytime. Certainly, it was much too faint for naked-eye daylight observations.

The second "legend" is a maximum tail length of 70 degrees. Again, there is no known reference to this in primary records. The greatest tail lengths that can be traced to original observations seem to be around 20–25 degrees, as noted above. The earliest known reference to a 70-degree tail (and most probably where the story started) is in Vsekhsvyatskij's *Catalog*, but he gives neither the observer's name nor any idea as to where he found the information. He does not even give a date other than "during December." But herein lies another problem... *December*! According to all reliable accounts, October was the month when the comet was brightest and its tail longest. By December, as we have just seen, the tail was down to just a few degrees. Even William Herschel, who earlier gave consistently long measurements, could trace it for just 5 degrees that month. Is it possible that there is a misprint in the *Catalog* (or in Vsekhs-vyatskij's unnamed and unlisted source) and that the real value is 7 degrees, or, perhaps, 70 min?

Either is possible. On the other hand, this very vague and unreferenced report may be totally spurious. Either way, the 70-degree December tail almost certainly joins the ghost hitchhiker and the cat that ate the chihuahua as just another urban legend of dubious provenance!

Comet C/1858 L1 (Donati)

This comet evolved into one of the most impressive of the nineteenth century and, in the opinion of many, was possibly the most beautiful (though not the most spectacular per se) ever seen, in virtue of what has been described as an "artistic" tail formation. It is among the better known comets beyond the strict confines of the astronomical community and has been the subject of several works of art, even giving its name to a racehorse in the 1970s!

In contrast with its later fame, it began very humbly.

On July 2, 1858, G. B. Donati of Florence discovered a telescopic comet about 3 min of arc in diameter and described as being "of near uniform bright-ness." At that time, there was no indication that this was anything other than a normal faint comet.

Although European astronomers became aware of the comet fairly early, word took a long time to reach the United States, where independent discoveries were made by H. M. Parkhurst on June 30 and Maria Mitchell on July 7. Throughout this time, the comet steadily brightened, but as July gave way to August, its drift ever deeper into evening twilight made it increasingly difficult to see. Still, on August 5, according to A. Reslhuber (Austria), it was "clearly visible" deep in twilight with around 2 min of coma visible. Ten days later, as it began pulling out of the deepest twilight zone, the comet was observed by J. C. Watson in Michigan and judged to be equal in brightness to a star of 4.5 magnitude.

The first person to see Donati's Comet with the eye alone appears to have been K. C. Bruhns (Berlin), who spied it without optical aid on August 28 and again on September 2, when he roughly estimated its brightness as in the 3–4 magnitude range. This more or less agrees with several other observers who placed it between second and third magnitude in early September, although it had brightened to 1.8 by the 15th and 1.4 by the 17th, according to Bruhns.

A tail began to develop in September, being just 50–60 min long on the 1st, 7 degrees on the 16th, and up to 25 degrees by month's end. Tail striae had evidently become visible as early as September 8, when S. H. Schwabe described it as being "striped and slightly curved toward the left." Although the description "striped" is a strange one, it is presumed that the stripes alluded to were most probably striae. Curvature of the tail is also implied by Reslhuber's characterization of it as "saber-shaped" on September 16.

The comet continued to brighten throughout September, reaching a persistent plateau of around 0 magnitude or brighter from the end of that month through the middle of October, according to extensive observations by Schmidt, Bond, Winnecke, Bruhns, and others. Between September 22 and October 8, several observers saw the comet in daylight using (what for that day were) moderately large telescopes.

The first daylight sighting was reported by Madler at Dorpat, who found it just 2 min before sunset on September 22 using the observatory's 24.4-cm (9.6-in.) Frauenhofer refractor. On the 17th and 19th he had used this telescope to observe the comet at sunrise and sunset, respectively, but had failed to find it when the Sun was actually above the horizon. At the time, the comet was probably shining at first magnitude or a little brighter. At the end of that same day, Bruhns failed to find it until 14 min after sunset with the 9.6-in. telescope at Berlin Observatory, the very telescope with which Neptune was discovered. Other early failed attempts were made by Schmidt on several occasions between September 22 and October 3 and by W. R. Dawes on September 30.

Success came for Bruhns on October 4, however, when he spied the comet 22 min prior to sunset. On that same late afternoon, Schmidt also found it just 12 min before the Sun went down and on the following afternoon, 13 min before. Earlier that day, Hodgson tried for it at 11 a.m. using a 16-cm (6.3-in.) refractor at 25 magnification without success, but an observer at Vienna (either Hornstein or Weiss) managed to find it just 6 min before the Sun set with a 16.3-cm (6.4-in.) refracting telescope. Also on September 5, G. Bond used the Harvard Observatory's 15-in. refractor to find the comet at 4 p.m., fully 1 h 35 min before sunset. This is the earliest in the afternoon that the comet was seen.

The comet was again seen prior to the Sun's setting on October 8, Bruhns finding it 2 min before and Dawes just 3 min before. According to the latter, "On applying my eye to the telescope at 5 h 20 m GMT, while the sun was shining brightly into the observatory, the comet was instantly seen in the center of the field. This was the only occasion on which I was able to detect it while the sun was above the horizon, though at the time I think it probable it might have been perceived at least ten minutes sooner. On two or three days I think it might have been perceived if the sky had been free from haze; but the effect of a very slight film of haze when acted upon by the sun is fatal to the visibility of such an object." (Anyone who has gone after a comet in daylight will agree with that last statement!) Dawes noted that the nucleus of the comet appeared to be shaped like a crescent when seen in daylight and had such a solid appearance that he thought it quite capable of creating the dark narrow lane that had become noticeable down the center of the tail and popularly called the "shadow of the nucleus." From his description of the comet as observed in daylight, it seems that only the brilliant nucleus was visible while the Sun was above the horizon. Dawes tried for another daylight observation on the 11th but failed to find it. After that date, neither Dawes nor anyone else reported any attempts to observe it during daylight hours.

Late September 1858 stands as a landmark occasion in the long history of comet observation by witnessing the first attempts at comet photography. Several photographs appear to have been made of this comet. One of the first was taken by English portrait artist W. Underwood, who managed to secure a 7-s exposure of the comet through an f/2.4 portrait lens on September 27. The attempt was partially successful, recording the bright central region of the coma and part of the comet's tail. The following day, a totally independent attempt was made by Bond at Harvard College Observatory using the f/15 telescope and an exposure time of 6 min. The tail was not recorded by this, the first ever, telescopic portrait of a comet. Only the bright central region was captured on the photograph.

The comet passed through perihelion (at 0.58 AU from the Sun) on September 30 and was nearest to Earth on October 10 at 0.54 AU. On the first day of the month of October, the main, curving dust tail was variously estimated as between 21 and 27 degrees, with a much fainter and very thin straight tail emerging from its convex side. The secondary tail was clearly a Type I plasma tail and was apparently too faint to be seen except under near perfect conditions. This tail increased in length from around 3 degrees on October 1 to 30 degrees on the 3rd, 32 degrees on the 4th, and 40 degrees 2 days later. A well-known lithograph showing the comet over the Conciergerie in Paris on October 5, thought to be by Mary Evans, shows two very thin and straight tails, one so long as to extend out of the field of view. The main tail on that evening is delicately curved, with the convex edge bright and well defined and the concave a lot fainter and more diffuse. The conspicuous circlet of Corona Borealis is visible to the left of the pair of straight tails, and the bright Arcturus shines very close to and a little to the right of the comet's head. From the relative appearance of the star and comet's head as depicted in the lithograph, the comet would appear to be clearly brighter than Arcturus. The magnitude of Arcturus is variously given as being between −0.06 and +0.04 (let us call it 0 magnitude), so if the lithograph is accurate, the comet would seem to have been around −1 as judged by the artist on that night. Naturally, we must beware of reading too much into a lithograph, but its accurate depiction of the stars and their brightness relative to one another, as well as the very realistic rendition of the comet itself, is worthy of note.

In the meantime, the main tail also increased in length, reaching around 40–43 degrees on October 11, just after the comet passed closest to Earth. At its widest, this dust tail was estimated as between 10 and 16 degrees. Striae, or maybe genuine synchronous bands, may also have been suggested on the 11th, when some observers noted that the tail was composed of multiple branches, although it apparently became featureless again during the following week.

Fig. 5.1 Donati's Comet on October 5, 1858, as drawn by C. Flammerion

Fig. 5.2 A rare stereo photograph of Donati's Comet, probably with darkroom help. © Stuart Schneider. Photographs from *Halley's Comet, Memories of 1910*, by S. Schneider, and *Wordcraft.net*

As the comet moved away from both Earth and the Sun, and observing conditions deteriorated, the apparent tail length diminished from 33 degrees on October 15 to 20 degrees the following night and to just 5 degrees on October 17! Sometime between October 7 and November 5, Chinese astronomers noted the presence of the comet in the form of a large "broom star."

During the latter half of October, the comet increasingly favored southern hemisphere observers as it continued its southward trek. By then, it had also faded considerably from its peak early in the month, as an estimate of magnitude 3.7 (October 25) by Wullerstorf and Muller indicated. Although some of this apparent fading was undoubtedly due to the comet's decreasing elevation for northern observers, it was obvious that its best display was now well and truly over. Even southern astronomers were reporting tail lengths of only 4 or 5 degrees after the Moon left the sky in late October.

It was becoming a difficult object to see with the naked eye in the second week of November, according to Callow, who saw it last without optical aid on November 11. This appears to have been the final naked-eye sighting. The same observer gave his final description of the tail on November 4 as "like a double bow" but unfortunately provided no estimate of its length. It was nevertheless unlikely to have been more than a couple of degrees at most. The comet was followed telescopically by C. W. Moesta in Santiago and W. Mann at the Cape of Good Hope until March 2 (Moesta) and March 4 (Mann), 1859.

An interesting feature of this comet, as seen telescopically, was the wealth of detail within its inner coma. The first indication of these structures came on September 15, when W. J. Forster noticed what he described as "a very distinct emanation" and a bright ray issuing from the nucleus. Another feature described as a "wisp or fan" was observed in a more or less sunward direction by A. J. G. F. von Auwers between October 1 and 14. This may have been a short anti-tail rather than a true "jet."

Closer to the visual nucleus, a hood was noted on October 1 by Reslhuber. Two days later, the same observer recorded that this had separated from the nucleus and seemed to be expanding out into the coma. At least six or seven of these hoods were seen during this period. They were brightest when first appearing very close to the nucleus but faded as they expanded outward into the coma. It seemed as though the nucleus was throwing off spurts of luminous material, rather like a spinning firework. The classical representation of these spirals was recorded in the beautiful drawings of Bond and Julius Schmidt. Alas, however, Bond paid little attention to recording the timing of these observations, greatly diminishing their usefulness in later years when the true nature of the inner coma structures became known.

It was Fred Whipple's icy conglomerate model of cometary nuclei, put forward in 1950, that provided a framework for solving the mystery of the spiraling jets seen in Donati and other comets. In a later analysis specifically referring to Donati and drawing heavily upon Schmidt's drawings, Whipple argued that these jets emanated from a single spot on the icy nucleus that was far more active than the average surface. Maybe there had been an explosive eruption of trapped subsurface gases during a previous return of the comet, or perhaps Donati had been hit by a meteorite at some time in the past. Or something else entirely might have caused a region of the less active surface to crack away and expose a deeper region of more volatile ices. Whatever the reason for this active spot, the spin of the nucleus about its axis meant that it spent time in daylight and time in darkness. It was when the Sun rose above the horizon of the nucleus (as seen from this spot) that this area of exposed ices became active. As the stream of gas and dust boiled from it, the nucleus continued to turn, and the rising plume swept back into a spiral-shaped arc. In a short while, the sunset over the horizon, activity ceased, and the plume (no longer fed with new material) detached from the nucleus and drifted off into the coma, swept back by solar radiation into the form of a hood. By determining the time elapsing between the onset of these regular spiral jets, and assuming that they really did issue from a single active spot, Whipple determined the period of rotation of the nucleus as 4.62 h. This is unusually fast for a comet (most determinations of the rotation period of other comets indicate days rather than hours) and is actually perilously close to the velocity at which a weak body such as the average comet nucleus starts flying apart!

Actually, F. A. T. Winnecke *did* report a secondary nucleus from October 7 until the 9th, though it is doubtful if this implied a true breaking of the nucleus. Certainly, there was no evidence of a serious breakup, and from the rapid acceleration of the secondary condensation, its brief duration and the apparent lack of any obvious flare in the comet's brightness, it seems that Winnecke's object, if a solid fragment at all, was only a very minor one. It may have been associated with the jets active at that time rather than anything solid breaking away from the nucleus itself.

It may be significant that no one other than Winnecke reported a separate nucleus per se, although there were several reports of the nucleus being elongated around that time. It seems likely that they were seeing the same thing as Winnecke, but for one reason or another were not resolving the double condensation into its separate components.

Donati's Comet was, as we have said, remarkably beautiful and left a deep impression on all who saw it. Nevertheless, just 3 years were to pass before Earth's skies were visited by another object that, although not possessing the graceful beauty of Donati, attained such brilliance as to make Donati's light seem almost feeble! I refer to the Great Comet of 1861, the American Civil War comet otherwise known as C/Tebbutt. Let us take a closer look at this remarkable object.

Comet C/1861 J1 (Tebbutt)

On the evening of May 13, 1861, John Tebbutt, Jr., of Windsor, New South Wales, was searching the western sky with a small marine telescope for possible comets, when he happened upon a faint nebulous object in the constellation of

Eridanus. He remarked that in his small instrument the object "appeared much diffused, and it was with the greatest difficulty that I measured its distance from three well known fixed stars." He also mentioned that it was "hardly distinguishable in the small telescope attached to the sextant" and very approximately estimated its brightness as "about the fifth magnitude." Not having a catalog of nebula, Tebbutt was unsure whether this object was really a comet or a nebula, and he therefore decided to keep watch on its position relative to the star Lacaille 1316. He chose this star because it was visible in the same field of view (of the marine telescope) as the comet, and by using this he avoided the difficulty of making frequent sextant measurements on what could have turned out to be a false alarm.

At first, it seemed as though it *was* a false alarm. The night of the 14th saw no obvious change in the object's position, and the following evening was cloudy. Tebbutt found it again on the 17th, but once again its position showed no obvious change. Clouds set in for the following three nights, but by then Tebbutt had all but given up hope of the object being a comet!

A fortuitous break in the clouds allowed one more attempt on the 21st, and, for the first time, a clear change in position was apparent. Cautiously, Tebbutt entered in his journal that he was "almost persuaded of the cometary nature of the nebula" and on that same evening sent off a letter to Rev W. Scott, the Government Astronomer at Sydney Observatory, announcing the possible discovery. The following evening, Tebbutt again saw the comet and was convinced that the movement suspected the previous night was real. He sent a second letter to Scott confirming his suspicions, and that same night Scott himself located the comet with the 8.3-cm (3.25-in.) equatorial refractor formerly housed at the Parramatta Observatory (which, by the way, was the telescope used to recover the first predicted return of Encke's Comet back in 1822). On May 27th, Scott noted that the comet had become "just visible after sunset to the naked eye."

Tebbutt continued to follow the comet, making positional measurements as accurately as he could with the sextant, and computed what he called "a rough approximate orbit" by June 15. This orbit was published in the *Sydney Morning Herald* newspaper (the chance of having the facts of a comet orbit published in the *Herald* these days is very remote!). The newspaper also published Tebbutt's prediction that the comet was destined to make a close approach to Earth on June 29, when our planet would pass "at no great distance from the extremity of its tail." He also suggested that the comet might be "visible in full daylight about that date." These predictions apparently caused no consternation among the public, a very different situation to what happened 49 years later when similar predictions were published concerning Halley's Comet!

Actually, Tebbutt had earlier suspected that the comet was receding from us and that it therefore seemed "extremely doubtful whether three suitable observations can be made in order to ensure even an approximate ... orbit." This pessimistic assessment was quickly challenged by J. J. Gleeson, who suggested (correctly as it turned out) that the comet's very slow apparent motion was due to an almost head-on approach to Earth and that it was consequently set to become a spectacular sight. In his letter to the *Herald* dated May 29, Gleeson actually referred to the comet as "this beautiful object," and it is to be wondered if this was intended as an actual description of the comet at that time or an anticipation of what he believed it would become. It seems doubtful that it was a current description, in view of Scott's statement that it was just visible with the naked eye only 2 days earlier!

By June 3, the comet was seen in morning twilight at Cape Town and estimated to be around magnitude 2 or 3 with some 3 degrees of tail visible. About 2 weeks later, however, the tail had become "larger than that of any comet seen in [the southern] hemisphere since the memorable one of 1843" according to a *Herald* correspondent writing only under the name of "Orion" on June 19. R. Ellery of Williamstown in Victoria noted that the tail was double on the morning of the 20th, with the western or main tail appearing some 40 degrees long and the eastern one about 5 degrees, diverging from the longer at an angle of 34 degrees. However, Scott apparently saw just 18 degrees of tail the following morning, possibly due to inferior observing conditions at his site.

On June 20, Tebbutt wrote in the *Herald* that "On the last two mornings, I have observed the tail to be divided into two branches which emanate from the main part of the tail at a distance of about six degrees from the head. The upper or western branch was the more distinct, and I could trace it to a distance of 42 degrees from the head. The tail, supposing it to point directly from the sun, will cross the earth's path about the 29th instant at a point which will be occupied by the Earth on the 2nd of July; so that it appears the Earth will have a narrow escape from being enveloped in the more diffused part of that appendage. The comet will be in conjunction with the sun about the beginning of next month, and will shortly afterwards become visible in the evenings in the north-west."

As Tebbutt and others predicted, the comet did indeed pass Earth at a distance of just 0.13 AU on June 30. Perihelion (at 0.82 AU) had occurred on June 12.

On the day of closest approach, Schmidt in Athens estimated the brightness of the comet's head as being "not as bright as Jupiter," yet soon afterward, he remarked that the total light from the comet's tail became so great as to cast shadows against a white wall!

In England, the brightness of the nucleus was judged to be intermediate between that of Venus and Jupiter (therefore about –3) by T. W. Webb. This same observer also noted that the comet had a "golden hue" at that time.

During the period of closest approach to Earth, the effect of forward scattering of sunlight by dust particles in the comet's coma and tail should have significantly enhanced its brightness. There are good reasons for thinking that this did happen.

For one thing, we have the "shadow casting tail" noted by Schmidt. Then, on the evening of June 30, the comet was "plainly visible at a quarter to 8 o'clock (during sunshine)" and there is even a report of one person momentarily mistaking the head of the comet for the rising Moon! All of this suggests a strong enhancement of brightness as the comet moved through forward-scattering geometry between Earth and the Sun.

There is a very strong possibility that Earth did actually encounter the tail on June 30. True, our planet did not actually bisect the Sun/comet line, but due to the typical curvature of Type II tails, that does not necessarily mean that we could not have passed through part of the dust tail. As evidence that this did happen, it was noted at the time that a number of auroral-like sky illuminations were reported from different parts of the world.

In England, astronomer J. remarked that "on Sunday evening . . . there was a peculiar phosphorescence or illumination of the sky . . . [which was] remarked by other observers as something unusual." Hind's observation was apparently confirmed by "Mr. Lowe of Highfield House," who also noted that "the sky . . . had a yellow auroral glare-like look, and the sun, though shining, gave but feeble light." He also added that the sense of dullness about the Sun was so strong that the vicar of the parish church had the pulpit candles lit at 7 o'clock even though the Sun was still well above the horizon. The comet was also said to have appeared much hazier that evening than on other nights, despite its great brilliance.

On the other side of the world, Tebbutt noted that auroras (if that's what they were) were widely reported in New South Wales at that time and that he had observed, on the evening of June 30, "a peculiar whitish light throughout the sky, but more particularly along the eastern horizon." In his opinion, "This could not have proceeded from the moon, but was probably caused by the diffused light of the comet's tail, which we are very near right now."

The head of the comet, being both active and passing so close to Earth, presented an interesting spectacle in telescopes. Surrounding the planet-like false nucleus, some six hoods or luminous veils could be discerned, the brightest nearest to the center and the faintest furthest away. The scene was graphically described by Webb as looking like "a number of light hazy clouds . . . floating around a miniature full moon." Observing the comet on June 30, R. Main of Radcliffe Observatory at Oxford described a "stream of light . . . from the upper apparent part of the nucleus [which] turned round towards the apparent west in the shape of a sickle. Another but fainter stream was seen on the apparent east side of the first stream, also turning round toward the west." Main's description of the brighter stream is almost exactly that of a similar phenomenon seen by Rob McNaught in Halley's Comet late in February 1986.

Observing the comet again on July 5, Main estimated the brightness of the nucleus as 1 and noted that the two streams of light were by then passing "symmetrically on each side of the nucleus." At the same time, C. H. F. Peters in New York, observing with a 34-cm. (13.8-in.) refractor, described "many fine jets streaming out of the nucleus, part of them recurving to the right, others to the left." In moderately large telescopes, the comet must have been a remarkable sight. A rather puzzling phenomenon noted in early July (notably on the 3rd by J. M. Gilliss of the Navel Observatory, Washington, DC) was the apparent "flashings or pulsations, closely resembling those of the aurora" within the central coma. Likewise Peters, four nights later, remarked that the inner envelope seemed to be "undulating" before his eyes. It is worth noting that, 21 years later, aurora-like pulsations were said to have flickered down the tail of Comet 1882 F1 (Wells), and similar accounts have been given of other comets over the years.

It is very difficult to see how these "aurora-like pulsations" can relate to physical phenomena within the tail itself, as the velocities of propagation would be prohibitively high. Raymond Lyttleton undoubtedly spoke for the vast majority of astronomers when he put the effect down to a trick of our atmosphere. Still, strange reports do come from credible observers from time to time. For example, one very experienced British observer watched a small nebulous knot travel down the plasma tail of Encke's comet in a matter of minutes. It is not too easy to blame this on a trick of the atmosphere! But back to the comet of 1861.

Following its great display in early July, the comet's brightness thereafter declined noticeably. From various reports, it also appears that the central false nucleus became fainter and more nebulous after the middle of the month, and the inner coma structures grew progressively less distinct.

On June 30 and the evenings following, tail lengths of between 90 and 122 degrees were routinely reported by experienced observers. Situated in the northern skies, the great tail was actually mistaken for an aurora by at least one observer whose view of the comet's head was blocked by a bank of clouds. According to H. Goldschmidt in Paris, the main tail was about 5 degrees wide some 20 degrees from the head on the night of July 3. At the time, he measured the total length as 75 degrees.

Although it was becoming obvious that the best of the spectacle had passed, estimates as great as 85 degrees for the length of the main tail were still being recorded on the 5th when its maximum width was judged to be as much as 10 degrees. A similar width was estimated for the secondary tail on that night. The length of this appendage was then traced to around 30 degrees. According to Main, the nucleus was about first magnitude and remained surrounded by considerable structure in the inner coma.

While traveling down the Shire River in Africa, famous missionary and explorer Dr. David Livingston independently sighted the comet on July 6. He noted its position in the constellation of Ursa Major and estimated the tail as 23 degrees long. The following evening, another explorer on the Shire River, John Kirk, wrote in his journal that, "This night we got the sight of a splendid comet in the Great Bear moving rapidly from the sun."

The main tail was still as long as 57 degrees on July 8, according to Schmidt, although estimates by other observers were generally more conservative and ranged from 31 degrees (K. Littrow) down to just 14 or 15 degrees as measured by Webb. The secondary tail seems to have disappeared by then. It is interesting to note that Webb remarked on the "streaked" appearance of the tail as viewed with the naked eye. It is not immediately obvious what these "streaks" could have been. They were, presumably, dust features, but without further information it is difficult to decide whether they were striae or genuine synchronous structures.

Another interesting remark by Webb concerned the color of the comet. He noted that it appeared white with the naked eye but tinged with a bluish-green coloration in the eyepiece of his telescope. Later, on July 15, he gave the color of the nucleus as "greenish-yellow." The greater intensity of light in the telescopic view would help to bring out any faint color present, but it may be instructive that the golden hue that he previously noted on June 30 apparently did not persist long into the following month. It is possible that this earlier coloration was at least partially due to the bright solar reflection spectrum enhanced by the forward scattering of the Sun's light. If the reflection spectrum was brighter then, more yellow light would have been added to the mix. The lower altitude of the comet might have added to this effect as well.

As the comet moved away from both Earth and the Sun, it grew dimmer and the tails decreased considerably in length. According to Schmidt, its magnitude had faded to about 3 on July 12. On that same night, Littrow estimated the tails as 30 and 21 degrees long. However, by the start of the second week of August, estimates of the comet's total light placed it close to fifth magnitude, with a false nucleus of magnitude 8 and a tail of just 2.5 degrees. By then, the brightness was probably similar to its discovery luster. The last known naked-eye detection of the comet was by E. Heis on the night of August 15. However, it remained visible in telescopes into the next year, the final observation apparently having been made by Winnecke and O. Struve at the Pulkovo Observatory in Russia on April 30, 1862. By that time it was probably not much brighter than 14th magnitude.

How did the Great Comet of 1861 rate as a spectacle in comparison with the other major comets of the nineteenth century? The person who could best answer this was probably Sir John Herschel, and, according to D. P. Todd, his verdict was that the 1861 comet "exceeded in brilliancy all other comets that he had ever seen, even those of 1811 and 1858." It is worth noting that Herschel also saw the Great Comet of 1843 (about which more will be said in the following chapter), although being in England at the time, his view of this object would not have been as clear as those further south.

The orbit of the Great Comet of 1861 is not typical of great comets in general. For one thing, the perihelion distance of 0.82 AU is somewhat large, though not excessively so. More importantly, the comet moves in a relatively short-period ellipse. The best-determined orbits appear to be those of A. Savitch and (especially) H. C. F. Kreutz, suggesting periods of 422 years and 409 years, respectively. For a great comet, these figures are really very small. The only comet of shorter period known to have been great on at least some returns is 1P/Halley!

This begs an interesting question. Has this comet been recorded at previous apparitions?

One may think that finding it at earlier times should have been easy, but it must be remembered that the great display provided in 1861 resulted to a large degree from its close passage of Earth. Had it appeared at a different time of year, and not come so close, the spectacle would have been much more subdued.

Still, it must be said that the comet is an intrinsically fairly bright one and should be found in old records. Given the uncertainty in the period, though, it probably would not have appeared exactly 409 years (or 422 years) prior to 1861, but these figures should at least give a ball-park value around which to search.

The best candidate for the previous return is the comet of May and June 1500. An alternative orbit, having little resemblance to that of 1861, gives perihelion at 1.11 AU on April 30, 1500, but the number and quality of observations of this object are not really sufficient to give too much confidence in this, and an orbit similar to that of Tebbutt's Comet can do equal justice to the information at hand. The period is a bit shorter than either Savitch or Kreutz calculated, but that is not a serious problem, and there are no other obvious candidates for the comet's previous return.

Before leaving our account of the 1861 object, we should mention that this was not the only great comet discovered by John Tebbutt. Almost exactly 20 years later, on May 22, 1881, he found a misty naked-eye spot low in the western twilight which, on being located in a telescope, turned out to be a close grouping of two stars and the head of a rather bright comet! This second Comet Tebbutt later evolved into a spectacular sight, especially after moving into northerly skies, and well deserves to be remembered as the Great Comet of 1881. It was not, however, equal to his earlier find and has not been listed among the greatest of the greats.

Comet C/1910 A1

The twentieth century began in grand fashion comet-wise, with the appearance in April 1901 of a great comet discovered by Viscara at Paysandu in Uruguay. For a time it displayed a multiple dust tail up to 15 degrees long and a faint ion tail reaching a full 45 degrees, but faded quickly as it moved away from the Sun and Earth. (Incidentally, the period of this comet was given in at least one publication as 39 years. Alas, no! This arose from a misreading of Vsekhsvyats-kij's *Catalog* where the period is given as 39,080 years, but, somehow, the comma became a decimal point, and this figure was subsequently printed as 39.08 years. If only!) This comet, though truly a great one to open the century, was not deemed sufficiently outstanding among these already remarkable objects to be officially included in the present list.

The next great comet of the twentieth century was a different matter. We refer, of course, to the Great Comet of 1910, more formally known as C/1910 A1 and popularly called "The Daylight Comet," "The January Comet," or, sometimes, "The Miner's Comet."

Fig. 5.3 The "Daylight Comet" 1910 A1 in late January 1910 (courtesy, Lowell Observatory)

The last appellation comes from the fact that the first people to see this visitor were three diamond miners in Transvaal, who spotted it with the naked eye deep in the morning twilight of January 12. No formal report was made, probably because they thought the authorities would already have known of it!

Then, three mornings later, some railway workers at Copier Junction in the Orange Free State also spied it and followed it for about 20 min with the naked eye. Thinking that it was the expected Halley's Comet (which was actually still too faint for naked-eye visibility), the station master reported the event to the *Leader* newspaper, and it was not until someone from that paper telephoned Transvaal Observatory director, R. T. A. Innes that news of the comet finally reached the astronomical community.

Innes and W. M. Worssell planned searches on the mornings of January 16 and 17 but were thwarted by clouds on the first occasion and very nearly on the second. However, just before sunrise on the second morning, a conveniently placed break in the cloud cover allowed the astronomers their confirmatory peek at the comet. Telegrams to this effect were then dispatched to the wider world, announcing that a bright new comet had been found.

There seems to have been some confusion initially. One telegram was sent out announcing that a great comet had been seen in the morning skies from

South Africa. Maybe it was the accent or some fault in the transmission, but the word "great" was heard as "Drake" by the receiver, and an announcement was made that "Drake's Comet" had been seen in South Africa!

In any case, the comet quickly became widely observed. Innes located it again with the naked eye in broad daylight about midday on the 17th just 4.5 degrees from the Sun's limb. To his unaided eyes, the comet appeared as a snowy-white object about one degree long, with a star-like head brighter than Venus at her best. Daytime observations were also reported from Vienna, Algiers, and Rome on the following day, and on the 19th the comet was followed by astronomers at the Santiago Observatory from 11:14 a.m. until 6:00 p.m., when it was located some 7 degrees east and 3.5 degrees north of the Sun. On that same day, further daylight sightings were made at Cambridge as well as at the Lick and Milan observatories.

Perihelion (0.13 AU) had occurred on January 17, and as the comet moved away from the Sun, it trekked northward and became very well placed for northern hemisphere observers. It was widely seen by the general public in both Europe and North America. Indeed, many people who in later years recalled having seen Halley's Comet in 1910 were found to be actually describing C/1910 A1!

After January 20, the comet was well placed in the evening twilight. Observing from Leeds in England on the evening of the 21st E. Hawkes wrote that

> The comet was picked up with the naked eye at 4 h 40 m, and was a gorgeous object. The picture presented in the western sky was one which will never be forgotten. A beautiful sunset had just taken place, and a long, low-lying strip of purple cloud stood out in bold relief against the glorious primrose of the sky behind. Away and to the right the horizon was topped by a perfectly cloudless sky of turquoise blue, which seemed to possess an unearthly light like that of the aurora borealis. High up in the south-west shone the planet Venus, resplendently brilliant, while below, and somewhat to the right, was the great comet itself, shining with a fiery golden light, its great tail stretching some seven or eight degrees above it. The tail was beautifully curved like a scimitar, and dwindled away into tenuity so that one could not see exactly where it ended. The nucleus was very bright, and seemed to vary. One minute it would be as bright as Mars in opposition, while at another it was estimated to be four times as bright. The tail, too, seemed to pulsate rapidly from the finest veil possible, to a sheaf of fiery mist.

Presumably the fluctuations in the comet's visibility of which Hawkes speaks in this beautifully descriptive passage relate to rapid changes in the clarity of the atmosphere rather than to any intrinsic property of the comet. It is interesting that he estimated the brightness of the comet to have been "four times as bright" as Mars at opposition. Mars is capable of reaching about –2.8 magnitude at very favorable oppositions, and four times (approximately 1.5 magnitudes) brighter than this would have made the comet as bright as Venus. With Venus close by in the skies, however, if these two objects were truly equal in luster, he probably would have said so and not compared it to the (remembered) brightness of Mars. It is more likely that Hawkes was thinking of more modest opposition magnitudes of Mars, maybe about –2, which would then have made the comet approximately –3.5 at the time. From his description, that does not seem an unrealistic estimate.

Two days later, the tail appeared as a larger scimitar in the evening sky; 25 degrees long and 5 degrees wide at the extremity, according to O. Lagerblad, who observed it from Karlshamn in Sweden.

Not everyone was so sanguine about the sudden appearance of this brilliant comet. Along an area of coastline in Portugal, many people gathered at sunset to watch it emerge from the darkening twilight. But they did not all come to enjoy the wonder of this beautiful object. Some at least came in fear! It was said that as the comet became visible, many within the gathered crowd "crossed themselves in fear." Even a newspaper report from January 27 carried an item in which the comet was blamed for wild winter weather being experienced in parts of Europe at the time (but it *was* January!) and predicted further dire consequences just around the corner.

Despite full moonlight, tail lengths of 18 degrees were recorded on the evening of January 26 and, after the Moon vacated the evening skies, estimates ranged from 30 to as great as 50 degrees in the last days of the month. An observer in the south of England rated the comet as the largest (although not necessarily the brightest at that time) that he had seen since Donati in 1858. In the opinion of this gentleman, all the intervening comets were but "poor little things" by comparison. Without wishing to detract in the slightest from either Donati or 1910 A1, one must suspect that he missed Tebbutt's 1861 comet at the time of its best display!

In the dark skies of late January, after the Moon had left the evening sky, some observers commented on an extended glow into which the tail appeared to merge and, although this was identified as zodiacal light, there may also have been some extension of the tail itself, perhaps similar to a modified version of the tremendous dust tail displayed by Comet McNaught in 2007. As we will see, this comet shared many similarities with 1910 A1, and we can only wonder at what images might have been secured if the digital cameras with which astrophotographers assaulted the latter comet had been available back in 1910!

The comet of 1910 was extremely rich in dust. A small dispersion spectrogram by F. Baldet displayed a purely continuous spectrum of reflected sunlight, not just from the photometric nucleus (where the continuum was especially intense) but also from the tail out to at least 8 degrees. Further spectrograms by W. H. Wright confirmed the very dusty nature of this comet, revealing its spectrum to be continuous up to at least 1 degree from the head. None of the usual cometary emissions were visible. However, Wright did find strong sodium emission from the head region, and H. F. Newall, observing the comet with a direct-vision prism inserted between the eye and eyepiece of a 63.5-cm (25-in.) telescope, traced the sodium emission beyond the head and into the tail itself. This is believed to be the first observation of a comet's "neutral sodium tail." Although observed in several later comets, this feature was not adequately analyzed until Comet Hale–Bopp in 1997. The presence of sodium, together with the strong presence of reflected sunlight, explains the description of the comet's color as yellow and even red, as noted by several observers.

The brilliance of the continuous reflection spectrum observed when the comet was near its brightest, undoubtedly owed a good deal to favorable forward-scattering geometry. As we will see in the case of our next entry, this phenomenon can so enhance the continuous spectrum of a very dusty comet as to totally drown out any emission lines that might be present.

This comet is considered one of the classical instances of dust tail striae. Descriptions at the time compare the tail to a curved horn, growing wider at the end and possessing an inhomogeneous structure with "isolated branches extending at considerable angles to the horn's axis." These striae were recorded on photographs, and some of this detail has been clarified by applying modern processing techniques to these old images. Once again, though, we can only wonder about the wealth of detail that digital cameras might have uncovered had they been invented then!

On January 27, a shorter, straight tail was prominent within the concavity of the larger one. Despite its straight form, this also appears to have been a dust feature and not a plasma tail, as one might suppose. In addition, a small anti-tail was apparent on a photograph taken by E. Barnard at the beginning of February. A weak plasma tail did, however, start to form as the sodium emission waned, and after January 26th the normal cometary emissions began showing up in the spectrum.

The comet faded as it retreated from both Earth and the Sun and was down to magnitude 6.1 on February 12, 8.2 on March 7, and 11 by April 10, according to estimates by G. van Biesbroeck. It was last observed as a very faint stain in the background sky on July 9, when probably no brighter than 14th magnitude.

Comet 1910 A1 appears to have been a new one in the sense that the 1910 apparition marked the first time that it dipped in from the Oort Cloud and closely approached the Sun. Thanks to the combined gravitational pull of the Sun and planets, however, it will now fall a little short of the Oort Cloud, and at some time in the distant future pay our region of space a second visit on its very elongated elliptical orbit. This will be no time soon, however. This comet will not grace our skies again for many thousands of years.

Comet C/1927 X1 (Skjellerup– Maristany)

Some readers may wonder why this comet was chosen for inclusion among the greatest of the greats, as its display was brief and it seems to have had the misfortune of being about as poorly placed as possible during its one and only return in recorded human history.

Nevertheless, the comet was a majestic sight for favorably placed observers. One person remembered it from childhood as "awe inspiring," and even the sedate science journal *Nature* referred to it at the time simply as "the great comet." In 1985, when comets had again become a popular topic as the astronomical world readied itself for the impending visit of Halley, an eyewitness of the 1927 comet contacted Queensland comet observer Terry Lovejoy with his reminiscences of that long-ago object. This observer, L. Thorpe, said that he had seen most of the bright comets of the century and judged 1927 X1 as the most spectacular of them all. It was, for instance, a far more impressive object than Bennett, the Great Comet of 1970. As he recalled, the comet appeared in the twilight as a brilliant golden object with a conspicuous tail perhaps 15 degrees long.

Reports such as this, of the twilight spectacle plus the great brilliance reached by this object near perihelion passage and its widespread visibility in broad daylight (both by eye alone and with optical aid) renders the comet deserving of a place among the greatest of the greats, even though its full splendor was much diminished by poor placement in the sky.

The comet was first noticed in the southern morning sky on November 27 and independently found by at least ten people. Among the sightings were some odd discovery tales!

As an exception to Lewis Swift's dictum that, "You can't find comets while lying in bed," one lady apparently did just that in 1927! Taking advantage of the warm summer nights by making up her bed on an open verandah, she woke up in the early hours to see the comet shining in the encroaching dawn!

Just how many independent and unrecognized discoveries took place during late November and early December is impossible to say. One of the first discoveries was by C. O'Connell in New Zealand on November 28, but the report was unfortunately delayed. The first report to reach the relevant authorities was that of John Francis Skjellerup (pronounced "shell-er-up"), formerly of South Africa but then living in Melbourne, Australia. Skjellerup's name is still heard among comet observers, as he had previously rediscovered a short-period object briefly observed at an earlier return by New Zealand observer J. Grigg. Periodic Comet Grigg–Skjellerup is one of the better known comets of very short period.

Skjellerup's discovery of C/1927 X1 was a lucky find in more ways than one. Apparently, he had no intention of seeking comets that morning and only ventured outside after being woken up by a strange noise. The noise, as it turned out, was something being knocked over by the cat! Finding a clear sky, Skjellerup decided to make the most of this rude awakening and quickly located the new comet!

The following morning, another independent discovery was made by Rhind at New Plymouth and yet another on December 6 by Edmundo Maristany at La Plata. Nowadays, Maristany's name is officially linked together with that of Skjellerup in the title of the comet, but for many years C/1927 X1 was simply known as "Comet Skjellerup" or "the Great Comet of 1927."

Given the number of separate discoveries, the comet was clearly already bright by the last days of November. At the time of Maristany's discovery, it was estimated as second magnitude and sported 3 degrees of tail in the twilight.

Because of its strong southerly declination, the comet was seen in both the morning and evening skies in early December, appearing low in the twilight to the naked eye as a bright glittering yellow object with a pale yellow tail widening toward the extremity. A photograph taken on the morning of December 8 revealed a condensation of about 48 s of arc in diameter and sprouting jets of material to both the west and east.

Because of the comet's location in bright twilight, astronomers had problems locating enough nearby stars to accurately fix its position, and for that reason determining its orbit was no easy task. An initial attempt at computing an orbit even suggested that it might be a return of the periodic Comet De Vico of 1846, but this hope was soon shattered as more positional measurements became available and the orbit better determined.

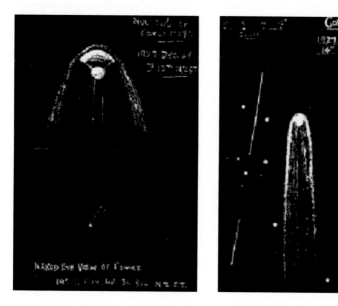

Fig. 5.4 Drawings by R. A. McIntosh of Comet Skjellerup-Maristany, December 5, 1927 (courtesy, W. Orchiston)

The comet's brightness increased rapidly following discovery, from around 3 when located by Skjellerup to second at the time of Maristany's discovery to first the following day.

Although C/1927 X1 is generally thought of as a southern comet, as it neared perihelion (0.18 AU on December 18) it was for a short time positioned north of the Sun and very close to it. Several northern discoveries were made at that time – on December 15 by P. R. Chidambara Aiyar at the Kodaikanal Solar Observatory in India (more will be said about this discovery below), by Warner at Hanover the following day (when 3 degrees of tail were noted), by a woman in a hiking party in the Sierra Madre mountains (who noticed the comet when the Sun was hidden by a mountain peak), and by a timber industry worker at Flagstaff, AZ, who noticed it in full daylight and reported it to the Lowell Observatory. (This seems to have been the first notification that the observatory received, but quickly put it to good use by making the first-ever observations of a comet at infrared wavelengths! We will say more about this shortly). The comet was also observed at Bergedorf on December 17 and 2 days later, C. Hoffmeister at Sonneberg and Muller in Potsdam estimated it as first magnitude, with the former tracing the tail for some 8 degrees. Following perihelion, the comet was once again south of the Sun and fading rapidly. Its brightness had dropped to 2.8 on December 21 and to 3 by Christmas Eve, according to George Van Biesbroeck at Yerkes Observatory. John Duncan of Wellesley College Observatory saw it on Christmas Day and again on December 27, but following this nothing appears to have been seen until around the time of New Year, when a long tail briefly appeared in the morning twilight.

Nevertheless, it was during those few days mid-month that the comet was truly remarkable and widely observed in broad daylight. Skjellerup himself was one of those fortunate enough to see the comet in broad daylight on December 15, when he viewed it within 2 degrees of the Sun through binoculars and a small telescope, the Sun being kept from view behind the edge of a house and chimney.

On the morning of that day Chidambara was taking spectroheliograms when his attention was drawn to a bright object close to the Sun. As there were several small clouds scudding by the Sun and shining bright silver in its rays, he initially supposed that it was one of these, but something was wrong. This cloud refused to move with the others! He quickly realized that this was no cloud, but a brilliant daylight comet, the brightest since the great sungrazer of 1882.

He followed the comet for 3 days telescopically and by unaided eye, in that time making several drawings of its changing appearance and perspective in relation to the Sun. Through a telescope on the morning of the following day, Chidambara described it as "a magnificent object," which "revealed a remarkable wealth of detail." He described the head as "a nebulous mass with an extremely bright nucleus from which emanated two luminous, curved arms," one of which displayed more curvature than the other. The tail was bifurcated and "enormous, not so much in length as in bulk." As morning gave way to afternoon, so the tail grew "in length as well as in volume" until (as he poetically expressed it) "I thought it was a huge giantess who had let loose the terrific glory of her copper-colored mass of tresses, and who was running away with her back turned against us." When he first saw the comet on December 15, Chidambara judged it as much superior in brilliance to the nearby Venus. A similar sentiment was expressed the following day by A. Adel, V. M. Slipher, and R. Ladenburg at Harvard.

It would seem that the comet became brighter than C/1910 A1. According to Baldet, it reached a magnitude of –6, while A. C. D. Crommelin thought that it may have become as bright as –10. These were not, however, estimates in the true sense of direct comparisons with other objects of known brightness. They seem rather to have been derived from various descriptions of the clarity with which the comet was seen near the Sun in full daylight and should therefore be taken with some caution. The –10 estimate is most probably a bit too bright. Yet, Slipher et al. described it as being "many times brighter than Venus," which would seem to make it *at least* –6 on December 16. Moreover, we must remember that when Chidambara first saw it and initially mistook it for a sunlit cloud, it was little more than 1 degree from the center of the Sun. For it to have been so readily noticed at that distance, –6 seems too conservative. Perhaps – 8 or even –9 would be more realistic! On any estimate, this comet became (albeit briefly) one of the most brilliant ever recorded, surpassed in recent centuries only by the great sungrazers of 1843, 1882, and 1965.

This intense brilliance was not solely due to the comet's small perihelion. In fact, the maximum brightness occurred prior to perihelion, and on the 18th (the date of perihelion itself) the comet's brightness was estimated as about 2 magnitudes fainter than Venus and intermediate between the brightness of Jupiter and Mercury (indicating –1 or a little brighter; certainly nothing like the brilliance noted on the 15th and 16th). Indeed, Slipher and colleagues noted that the comet was clearly fainter in the afternoon of December 16th than it had been on the morning of the same day, and that on the following day, a telescope was required to see it in daylight. The final daylight observation made by this team was on December 19, although Graff in Vienna saw it through a telescope during daytime hours on the 20th, when he also noted parabolic envelopes within the head. The same observer likewise managed to see the head and a short section of tail through a small telescope just 5 min before sunrise on the 21st.

This apparently strange behavior has been examined in recent years by Dr. Joseph Marcus as part of a study of forward scattering of sunlight by dust in comets passing between Earth and the Sun. Sure enough, the Earth/Sun/comet geometry was such that the greatest forward-scatter effect should have occurred on December 15, just as the records of daytime visibility suggest. The effect should have noticeably declined by perihelion, resulting in a rapid fading of the comet as seen from Earth, even though it was actually brightening intrinsically because of its continued approach to the Sun.

As noted also for C/1910 A1 under similar circumstances, the effect of forward scattering so enhanced the brightness of the comet's continuous spectrum as to completely drown out any emission lines that may have been present. Slipher et al. write that on the morning of December 16 "only a strong continuous spectrum ... of the solar type" was noted at both high and low dispersion. However, "The next day, it was seen that the dark D-lines of sodium [i.e., absorption lines in the solar continuum] were bordered with faint bright D-lines [i.e., emission lines] of the comet. On December 18 these had strengthened until they were a very conspicuous pair of bright lines of equal intensity superimposed upon the sky spectrum. The following day they were still brighter.... The emitting sodium spread to a considerable distance from the nucleus and was most intense on December 19, the last day of observation."

At Lowell Observatory in Flagstaff, AZ, Carl O. Lampland measured the comet's infrared radiation in broad daylight from December 16 to December 19, using a radiometer attached to the observatory's 102-cm (40-in.) reflector telescope. This marked the first series of infrared observations of a comet, something that would not be repeated again until the apparition of Ikeya–Seki in 1965. These observations revealed an enhancement corresponding to for-ward-scatter geometry.

In deep twilight on December 19, Hoffmeister at Sonneberg and R. Muller at Potsdam traced the tail for 8 degrees in binoculars, as already mentioned, but around the time of the New Year, a far longer tail was observed rising out of the dawn from a head still too deep in twilight to be discerned. Lengths of up to 35 degrees were reported between December 29 and January 3. This tail was described as being slightly curved and, judging by its length and the poor observing geometry at the time, was almost certainly a Type II dust tail seen neither face on nor edge on but close enough to the line of sight to give it a fairly narrow profile. An ion tail, pointing almost directly away from the Sun, would have been very foreshortened and could not have achieved the lengths reported. According to Hoffmeister, the tail was 2 degrees wide on December 30.

It is interesting to note that on December 29, Hoffmeister described the 35-degree tail as "slightly twisted." Exactly what that meant is unclear, but it may be worth noting that a similar description was given to the narrow dust tail of Ikeya–Seki in 1965. The "twists" appearing in the tail of that comet were actually short, bright, striae, giving it an almost corkscrew-like appearance. Could Skjellerup–Maristany have displayed a similar striae pattern? This is a possibility worth considering.

Although not very well observed thanks to the poor placement of the comet, 1927 X1 displayed jets and envelopes within its inner coma not unlike those of comets such as Donati and Coggia of 1874. Indeed, the comet was compared to this latter object, both for the parabolic envelopes within its coma and for the dark band (sometimes inaccurately referred to as the shadow of the nucleus) that was observed at times to run down the middle of its tail. Unlike Coggia's, though, 1927 X1 made no close approach to Earth; we can only imagine the spectacle that it would have provided if it had!

The observations of the tail on January 3, 1928, were the last naked-eye sightings of this comet. Its magnitude was estimated as 9 in early February and 10 by the end of that month, when the coma was measured as just 1 min of arc in diameter. The final observation was made on April 28 from Johannesburg, by which time the comet had faded to near 14th magnitude.

Unlike C/1910 A1, Skjellerup–Maristany was not making its maiden voy-age to the Sun in 1927, although it is a comet of very long period; in the vicinity of 37,000 years, in fact. Maybe it will be seen to better advantage next time around!

Comet C/1975 V1 (West)

The remaining comets, together with Ikeya–Seki, which we shall meet in Chapter 6, are rather different from the preceding ones, as they are all objects that this author has personally observed. Of course, there is always the possibility of personal bias when writing about something that has been personally experienced.

There is a story about two professors of history talking 1 day about their respective specialties. One commented how fortunate he was that his special field of study was prior to 500 years ago. "I am even more fortunate," said the second professor. "My specialty is older than 1,000 years!" The point being, of course, the further back in time one deals, the less emotional involvement one is likely to feel with the subject and the more objective one is likely to be.

A critic might think that the inclusion of five of history's greatest comets within one's own lifetime is a clear case of bias. However, the comets Ikeya–Seki (1965), West (1976), Hyakutake (1996), Hale–Bopp (1997), and McNaught (2007) would surely be included had they arrived in past ages, and the fact that they just happened to appear recently is nothing more than a fortunate coincidence for us. In any more or less random sequence, there will be periods of higher than normal concentration. In the history of comets, you could note especially the 1840s to the 1860s, the 1880s, and the 1960s, where there was a "bull market" in bright objects. It seems that the 50 years between 1957 and 2007 was also "bullish" with respect to top-grade great comets, as well as several others that (for reasons explained earlier) did not quite make our present list.

Of the five comets mentioned above, we leave Ikeya–Seki to Chapter 6 and begin with Comet West, the Great Comet of 1976. As we will see, there is a sense in which this was an accidental great comet, or one that had greatness thrust upon it by an unforeseen and unpredictable event, but more of this later.

The discovery of this comet appeared deceptively routine. Images of the 14–15 magnitude objects were found by Richard M. West (Geneva) on photographs taken by Guido Pizarro using the 100-cm (39-in.) Schmidt camera at the European Southern Observatory at La Silla (Chile) as part of the "quick blue survey" associated with the southern Sky Atlas. The photographs had actually been taken on September 24, but West's investigation of earlier photographs also uncovered previous images of a comet as far back as August 10 on plates taken by Oscar Pizarro and again on August 13 on plates by Guido Pizarro. Total estimated magnitude was 16–17 on the first date and 16 on the second. At first, it was not absolutely clear if one comet or two had been found, but preliminary calculations by B. G. Marsden on the assumption that the August and September comets were one and the same produced a realistic-looking orbit. Moreover, if further proof was needed, this orbit led to confirmatory photographs at La Silla on November 8–11. In August, the comet had appeared with a coma about 2 or 3 s of arc in diameter and a trace of tail to the north. On the November photographs, it appeared to be some 20 s of arc across, well condensed and of the 13th magnitude.

From the computed orbit, it seemed that the comet was bound for quite a small perihelion distance (0.20 AU) on February 25 and would likely emerge from the morning twilight in early March 1976, as a naked-eye object more favorably placed for northern observers. Reckoning from the available magnitude estimates, brightness in the mid 3 s seemed likely at the beginning of the second week of March (when the comet's elongation from the Sun would be approaching 30 degrees) and around 5 by mid-month, by which time it would be very favorably placed in the early morning skies. These were interesting, if not profoundly exciting, prospects.

The first visual observation came on November 25 by the Reverend Leo Boethin in the Philippines, himself a discoverer of a periodic comet earlier that year. Boethin estimated the brightness of the comet as 12.7 at that time, with a coma some 4 min of arc in diameter and no sign of a tail. However, a broad tail 1 min of arc long was photographed on December 6 by C. Torres at Cerro El Roble. For a brief time in early December, the comet became accessible to northern observers and was recorded by T. Seki in Japan on December 1 (when estimated as magnitude 12.5) and by H. Giclas at Lowell Observatory 5 days later. Following this brief preliminary showing, it became badly placed for locations north of the equator, while remaining accessible in the early evening sky for observers in the southern hemisphere.

By December 23, the comet had reached a magnitude of 9.5 according to Albert Jones in New Zealand, who observed it with a 7.8-cm (3-in.) refractor. On December 30, the same observer found that its brightness had increased to magnitude 8.7. This author saw it faintly on January 1, 2, and 4 through 20 × 65 binoculars and thought it between magnitude 9.5 and 10; apparently an underestimate. On the other hand, M. Owada, observing with a 6.5-cm (2.6-in.) refractor at a magnification of just 12 from somewhere in Australia, estimated it as between magnitude 7.5 and 8 on January 7 (probably a little optimistic). J. Cooper (New Zealand) made it 7.5 on January 20, and his fellow countryman D. Goodman gave it 5.8 and 5.6 on January 30 and 31, respectively, as seen through 10 × 50 binoculars. These magnitude estimates showed the comet to be running up to 2 magnitudes ahead of predictions, although there were indications that the brightness rise was slowing and, though prospects for March seemed brighter than initially predicted, few if any people were thinking great comet at that time.

This is actually a point worth making in view of what is sometimes written about the so-called failure of the astronomical community to make public the prospects of a great morning comet in March 1976. We sometimes read that this silence was attributed to astronomers not wanting to go public with bright comet predictions following the "fiasco" of Comet Kohoutek in 1973/4. Actually, Kohoutek was not quite the fizzler that is widely thought and did become relatively bright with a long tail, although this was never very intense with the naked eye. But the real reason for astronomers' silence regarding West is simply that they were as surprised as anybody else at the comet's performance! Even in February, some astronomers were saying that the comet would probably not be a very impressive sight when it emerged from the dawn the following month.

On February 12, it seemed to be about third magnitude and sported a faint tail in deep twilight. Two evenings later, the yellowish object looked almost stellar except for a fan-shaped tail. It would have been very impressive indeed had the sky been darker. On the 17th this author found it among high cirrus clouds close to the horizon.

The first of a series of surprises came on February 22, when Richard Keen and G. Emerson in Colorado, C. Townsend in California, and, in New Mexico, Alan Hale (the same Alan Hale who in future years became famous as one of the co-discoverers of Comet Hale–Bopp) found the comet gleaming brightly at magnitude –1.

The second surprise occurred on the day of perihelion (February 25) when P. Collins and S. O'Meara used 6 cm (2.4-in.) and 23 cm (9-in.) refractors at Harvard Observatory to observe the comet in broad daylight. Through the larger telescope, O'Meara observed interesting details in the comet's head and tail half an hour before sunset. A drawing of the comet as seen at the time showed a more or less parabolic envelope with a bright sector on the sunward side and a slightly offset, brilliant, nucleus. Behind the nucleus there appeared to be a shadow running down the middle of the broad tail. A curious feature, looking like a small and distorted parabolic string, looped around the nucleus and extended its open end toward the Sun. Both edges of this loop cut across the parabolic coma envelope and extended sunward from the rest of the coma, looking rather like two horns projecting toward the Sun through the envelope of the coma. Collins and O'Meara estimated the comet's brightness as magnitude –2 at the time of these daylight observations.

That same day at Brooks Observatory in Stormville, NY, veteran comet observer John Bortle located the comet in full daylight with his 32-cm (12.6-in.) reflector, describing it as "a brilliant almost stellar object" 10–15 s of arc in diameter. It was white in color and reminded Bortle of a planet observed under conditions of poor seeing. He also noted the parabolic sunward fan and a tail some 30–40 min of arc long extending away from the Sun. Some 35 min later, he found the comet in 10 × 50 binoculars and estimated it to be –3 "brilliant like the planet Venus but with a bright 15-second tail!" Just 7 min before the Sun disappeared over the horizon, Bortle managed to spot it with his unaided eyes at a mere 7 degrees from the Sun. He again saw the comet at noon on February 27, both with the 32-cm reflector and 15 × 80 binoculars, and estimated the magnitude in the latter instrument as –2.4, using Venus and Mercury as comparison objects. A tail of some 2 or 3 min of arc could still be traced against the daytime sky.

This sudden surge in brilliance was partially due to favorable forward-scattering geometry, but there is evidence that this was in turn enhanced by a release of dust from the nucleus. The reason for this soon became apparent, as we shall shortly see.

In the meantime, the stage was set for a magnificent early morning display. This began slowly, with the comet appearing very deep in the dawn on March 1 as an object of magnitude –1 or thereabouts and sporting a tail visible for at least 2 degrees. Closest approach to Earth (0.8 AU) occurred on the 4th, and as the comet pulled out further from the Sun, its true grandeur was revealed in all its glory. In the words of Gil Wood, who negotiated the drive to the summit of Mt. Pinos, CA on March 7, "I glanced out the car window and there it was: Comet West, a fantastic fountain of light, flickering plainly through passing groves of trees. The head of the comet was too low to be seen over the embankment ... but the tail fanned out over the Cygnus Milky Way." Once the (false) nucleus was in view, he described it as "[burning] with the yellow brilliance of Venus inside the blue-white spray of coma." Just 3 days earlier, Gunnar Glitscher in Germany estimated the tail as 20 degrees long and having a surface brightness exceeding that of the star clouds in Cygnus. Tail lengths as great as 30–40 degrees were measured on the 7th by Dan Green, NC. This refers to the main dust tail. A shorter (about 10 degrees) plasma tail was also visible at the time, as well as a broad and faint fan of dust extending leftward from the bright edge of the main dust tail. Both the main dust tail and this faint fan were not homogeneous but were clearly comprised of several bands, sharper in the main tail than in the leftward feature.

Fig. 5.5 Comet West from The Netherlands on March 5, 1976. ©2002 R. J. Bouma, used with permission

According to a detailed study by Zdenek Sekanina and John Farrell, the features in the main dust tail fell into two distinct classes – genuine synchrones radiating outward from the nucleus, and striae leading back from the bright edge of the tail and resulting from the disruption of myriads of large particles ejected from the comet's nucleus. These two types of feature often intersected each other at angles of from 5 to 10 degrees, making interesting bifurcated spikes within the tail.

As the comet moved away from Earth and the Sun, both its brightness and tail length declined. According to Charles Morris, it was magnitude 1.4 on March 6, around the time that tail lengths of up to 40 degrees were being measured. Its brightness had declined to 2.3 by the 12th according to M. Mayo, and to 3.1 on the 18th as estimated by John Bortle. The dust tail had faded significantly by March 13, as the plasma tail became the more prominent appendage. The greatest act of the show was almost over.

Nevertheless, the comet remained a very impressive binocular object until well into April as it climbed even higher into the morning sky, and the tail could still be seen with the naked-eye on April 2, when the comet was roughly estimated as magnitude 4.5. The comet itself was still marginally visible with the naked eye on April 11, and tail lengths of up to a degree were plainly visible in 20 × 65 binoculars as late as May 25 and even faintly until the end of June, at which time the total magnitude had declined to around 7.5, according to binocular observers. The final visual observation appears to have been made by John Bortle on August 25, when he estimated the comet's magnitude as 11.

It was mentioned earlier that there appears to have been a surge in dust production just before the comet reached perihelion. The first clue as to what happened came in early March, when the nucleus started taking on an elongated appearance. By 5th it had became clearly double. Splitting into two components is not all that rare among comets, especially those that venture close to the Sun, but when two further nuclei appeared on the 13th astronomers began to realize that something quite unusual had taken place. The four nuclei (the main one and its three siblings) initially assumed a formation resembling a miniature version of the famous Trapezium in Orion, but this was soon disrupted by the rapid acceleration of the smallest member of the group. At the time of West's appearance, only two previous comets had fragmented into so many pieces – the Great Comet of 1882 and P/Brooks in 1889. The first passed very close to the Sun and the second very close to Jupiter. By contrast, Comet West did not pass exceptionally close to anything. Since 1976, the Jupiter-impacting periodic comet Shoemaker–Levy and 73P/Schwassmann–Wachmann have fragmented into more pieces, the first due to an extremely close passage of Jupiter and the second following (or accompanying) an enormous brightness outburst in 1995. This latter is, perhaps, closest in nature to the disruption of Comet West.

The West fragments were given the designations of A (the "main comet"), B, C, and D. Of these, C was clearly the smallest, as indicated by its rapid acceleration due to the rocket effect of escaping gases, and its rather rapid fading. It was last seen on March 27 and probably evaporated away to nothing shortly thereafter. The others were still visible on the final definite photographic image of the comet taken by C. Shao on September 25. Just 6 days earlier, Elizabeth Roemer estimated the brightness of the condensations as 19.3 for A, 20.1 for B, and 20.3 for D. Nucleus A may have been photographed again as late as October 23, although some doubt surrounds this observation.

The disruption of Comet West was studied by Sekanina and independently by L. Kresak in 1981. According to Kresak, D was the first to split away from the main mass on February 19, followed by B on February 27, and by the small C on March 6. The date of the initial split fits nicely with the first evidence of enhanced brightness and dust production from the comet. A major split of this magnitude inevitably exposes new reserves of fresh ice to solar heat and nearly always results in a surge of activity within a comet. A series of major splits, as happened to West, adds surge upon surge, and it was surely the cumulative effect of these disruptions that catapulted the comet from being a merely bright object into the class of the greatest of the greats. As Sekanina phrased it, "all the spectacular features displayed by Comet West are actually deadly scars and wounds inflicted on its not very cohesive body." As a lamp blazes brightest before going out, so Comet West's great performance appears to be a signal that its end is approaching. When last it passed through the inner Solar System, in prehistoric times, it was unlikely to have been as spectacular nor is it likely to be a great comet the next time it appears.

But even though it may be dying and have few returns left, its end is hardly imminent in terms of earthly years. West approached the Sun in 1976 on an elliptical orbit that suggested a period of around 300,000 years. Its aphelion distance fell short of the Oort Cloud, indicating that this is a comet that has been around the circuit at least once before. But because of the gravitational boost received from the planets during its recent appearance, this elliptical orbit has been so greatly stretched that the new aphelion actually falls within the Oort Cloud itself. Consequently, when it next returns (about a million years from now), it will approach on an orbit indistinguishable from that of a first-time comet. What could be its final approach to the Sun may masquerade as its first!

Comet C/1996 B2 (Hyakutake)

Very early in 1996, this author was talking to another comet enthusiast about (what else at that time!) the approach of the giant comet C/1995 O1 (Hale–Bopp), when he shared with me the following experience. The previous Sunday evening, walking home from church, he looked up into a clear night sky and thought to himself "Wow! Out there is a really *big* comet coming our way!" Then, at that very moment, the thought just popped into his mind "Yes, and there's another big one out there as well!" This was told to me simply as a slightly odd happening without any thought of "prophecy" or anything of that nature. At the time, Hale–Bopp was the only significant comet in the offing, and nobody expected another to turn up in the immediate future. Yet, within a week or two of our conversation, the (almost) unthinkable had happened. Another object with possible great comet potential had indeed been found!

Coincidence or prophecy? You be the judge!

The comet was found visually on January 31, 1996, by Japan's Yuji Hyakutake, who tracked it down using a pair of 25×150 binoculars. Coincidentally, this same observer had only recently (on Christmas Day, in fact) discovered another comet very near the same position. The first Comet Hyakutake – designated C/1995 Y1 – was in no way related to the second object, having an entirely dissimilar orbit, and it never reached naked-eye visibility.

The second did not look especially noteworthy at discovery, either. Hyakutake estimated its brightness as about magnitude 11, although this appears to have been a little conservative. Later, a prediscovery photograph from January 1 was found showing the comet at magnitude 13.3.

On February 1, this author picked it up using 25×100 binoculars and estimated its magnitude as 8.9, noting that it appeared to be rather strongly condensed and that its visibility was enhanced somewhat when viewed through a Swan Band filter. This latter fact implied that a goodly portion of the comet's light was from gases rather than reflecting off particles of dust in the coma. This in turn implied that the comet was not a very dusty one.

Within a few days, a sufficient number of precise positions had been determined at a number of stations for an orbit to be calculated. This provided the first surprise. According to the initial orbit, the comet was destined to pass close to both the Sun and Earth and ride high in the sky when nearest to us. Judging by the early magnitude estimates, it also appeared to be a moderately bright comet intrinsically, holding out the promise of an impressive display. Indeed, if it was to develop a significant tail, it could evolve into a truly magnificent sight. Updated orbital computations confirmed these prospects.

The comet was headed for a perihelion of only 0.23 AU on May 1, but even more exciting was the predicted approach to a mere 0.10 AU (just 15 million kilometers or 9.5 million miles) of Earth on March 25. Moreover, at that time the comet would be high in the sky for northern hemisphere observers. Indeed, it would be circumpolar, passing just 3.5 degrees from the north celestial pole little more than a day later. The comet's brightness at that time was predicted to be around 0 magnitude, offering an astronomical phenomenon rarely witnessed – a very bright comet in a dark night sky far from any twilight glow.

Hyakutake brightened steadily through February, being about magnitude 7.5 in the middle of the month and 7 on the 25th, when a faint westward pointing tail became visible in large binoculars. The first naked-eye sighting – under very favorable skies – was reported by Terry Lovejoy in Queensland, Australia, on February 27.

On March 11, this author estimated the comet as magnitude 4.6 with the naked eye under conditions that by no stretch of the imagination could be called ideal. That was about as bright as Halley's Comet would have been under similar circumstances, and measurements of the comet's water production at that time did indeed yield similar results to those of Halley. The stage appeared set for a spectacular show!

March 16 saw the comet glowing at magnitude 2.9 and sporting a short tail visible with the naked eye. Twenty-four hours later, it had brightened by two tenths of a magnitude, and the naked-eye tail extended for at least 3–4 degrees, and maybe as long as 10 degrees when viewed with averted vision.

The comet's northward journey carried it across the celestial equator on March 20, on which date its magnitude was estimated to stand at 1.7 and the tail could clearly be seen, albeit rather faintly, with the unaided eye. The comet had now become a very impressive sight. Its daily motion was also speeding up as it drew nearer to its passage of our planet. Like a locomotive approaching the station, until that time it had been coming essentially straight toward us, but by the end of March's third week, our line of sight was starting to become more tangential to its path, ready for its rush through the station on the way to Destination Perihelion.

The following night, this author had the opportunity of seeing the comet under very dark and clear rural skies, and the experience was one not easily forgotten. Late in the evening, the comet rose north and east like an enormous ball of luminous mist. It is tempting to describe it as looking like the ghost of the full Moon, except that with a coma diameter of 1 degree and a total magnitude estimated as 0.7, it simply looked too large and (in a manner of speaking) too *solid* to be called a ghost. The central condensation shone like an intense but fuzzy star in the eye of this cyclone of light, and two tails – a faint dust tail and a plasma tail that had intensified greatly during the preceding 24 h – extended upward into the sky. The two tails were estimated as 8 and 7 degrees, respectively, but about 2.6 degrees from the center of the head, the plasma tail split into a Y shape, with the southern component seeming to have disconnected from the main, northern, one. Close to the point of disconnection, the southern segment was remarkably bright, brighter indeed than any part of the main tail. Later in the night, when the comet had risen higher into the sky, the main tail appeared much longer. By 2 a.m. local time, when the comet was riding high in a very clear sky, the main tail was traced for possibly 33 degrees with the unaided eye.

My best view of the comet came the following evening from Darby's Falls Observatory, near Cowra in rural New South Wales. Arriving at the observatory around 10.30 p.m. local time, a long searchlight beam rising from beneath the northeastern horizon immediately captured the attention. This was, of course, no searchlight. Soon afterwards, the comet's head hove into view from behind a range of hills, and as the spectacle rose higher into the night, its full grandeur became more and more apparent. Now shining at magnitude 0.3, the comet outshone all but the brightest stars and its reflection gleamed in the windows of vehicles in the observatory's parking lot!

The tail was intense for about 30 degrees and more faintly visible out to some 45 or 50 degrees with the naked eye. I could trace it up to the Gegenschein into which it merged. At that distance from the head, the tail was quite a deal fainter than the Gegenschein, and I failed to pick it up again on the other side. There were reports, however, that some observers did manage that feat and traced it for several degrees beyond the Gegenschein's far side.

An interesting feature of the central condensation that night was a jet-like feature extending tail-ward from the false nucleus. Actually, it looked not so much like a jet in the true sense of the word as a plume of material (dust?) blown off the nucleus and swept back in the direction of the tail. The feature itself was quite conspicuous, easily visible in 25 × 100 binoculars and striking in the observatory's 50-cm (20-in.) reflector. In this instrument, it looked for all the world like a miniature comet enclosed within the core of a great one; a "comet within a comet."

Sweeping rapidly northward during the following days, the comet became an incredible spectacle for the northern hemisphere. There appears to have been a shallow and prolonged brightness outburst (although calling it an "outburst" is probably being too generous) around the time the comet passed near Earth's orbit, contributing further to its luster. On March 23 and for a few days following, several knots of material were seen moving away from the nucleus. These took the form of transitory mini-comets and were probably small icy fragments detaching from the surface of the main body. At no time, though, did the comet appear split in the true sense of the word. Certainly, there was never anything remotely resembling the multiple breakup of Comet West. On the night of March 23–24, the comet passed directly overhead for those near + 40 degrees latitude!

Fig. 5.6 Comet Hyakutake, March 24, 1996 (courtesy, Terry Lovejoy)

Fig. 5.7 The long plasma tail of Comet Hyakutake is seen projecting above the horizon on March 24, 1996 (courtesy, Terry Lovejoy)

Fig. 5.8 Comet Hyakutake on May 5, 1996, showing both plasma and dust tail features (courtesy, Case Rijsdijk, SAAO)

Fig. 5.9 The near-nucleus region of Comet Hyakutake imaged by the NASA Hubble Space Telescope as the comet passed close to Earth on March 25, 1996. The image at left shows dust being produced from the sunward-facing hemisphere of the nucleus. Three small fragments (each producing their own tails) are visible on the original photograph at upper left in this image. The image at bottom right shows a region of just 760 km (470 miles) centered on the nucleus (located at the tip of the bright jet). The image at top right shows pieces of the nucleus that were also detected from ground-based telescopes on March 24. Credit: H. A. Weaver, HST Comet Hyakutake Observing team, and NASA

At the time of closest approach to Earth, Hyakutake shone at around 0 magnitude and the tail extended at least half way across the sky. Some observers estimated the tail on that night and subsequent evenings to stretch for at least 100 degrees. The longest published estimate seems to have been 118 degrees, although several even longer ones were apparently reported.

These estimates became the subject of lively and, unfortunately, at times acrimonious debate. The problem was not so much the extreme lengths themselves, as tail lengths in excess of 100 degrees do occasionally occur. The big problem concerned the observational geometry at the time these estimates were being reported. In theory, tail lengths of these dimensions should have been impossible. A little elementary trigonometry shows that for a straight tail directed precisely away from the Sun, angular lengths of the order of those reported would require a real length greater than infinity – a logical as well as a physical impossibility. To be sure, in the real universe even plasma tails need not be perfectly straight, and there is nearly always some divergence from strictly anti-solar orientation. But any departures from the ideal would need to be pretty extreme to be of much help in this instance, and many astronomers, therefore, dismissed the longest estimates as illusory.

This would probably be the safest position to take on the matter if the extreme lengths were all reported by novice observers. But that was not so. Very experienced comet observers, such as Charles Morris, James Scotti, and David Levy all reported seeing 100 degrees of tail or longer around the time of closest approach, and the opinions of people with their experience need to be taken seriously.

Incidentally, at the Comet Conference held at Cambridge, England, in 1999, this issue was raised in the presence of Charles Morris, who reluctantly conceded that he *may* have been mistaken in his measurements. But no concessions were made by Scotti or Levy, and the possibility that so many experienced comet observers all suffered from the same optical illusion at the same time is in itself a mystery crying out for explanation!

It may be of relevance to note that a major disconnection event in the plasma tail was noted between March 24 and 25, straddling the time of closest approach. It is not unusual for plasma tails to grossly distort when such events are taking place, and kinks at right angles to the extended Sun-comet line are not unknown. A major kink bringing the tail toward Earth along the line of sight might be capable of explaining these excessive tail-length estimates.

Moreover, it was later discovered that the spacecraft Ulysses passed through a region of plasma at 3.73 AU from the Sun on May 1, and this was subsequently identified as the extended plasma tail of Comet Hyakutake. This implied that the tail was then at least 3.8 AU long and also strongly curved at great distances from the Sun. This finding has been put forward by some as supporting the extreme tail measurements, although skeptics correctly point out that what Ulysses encountered was not the *visible* tail but a greatly diffused plasma extending far beyond this feature. Whether the Ulysses results add any support to the observations of Morris and the others is likely to remain a moot point.

Whatever the true length of tail around the time of closest approach, the comet was certainly an incredible sight. Charles Morris admitted to being so overwhelmed that he just stood staring at "The most unbelievable sight I've ever seen in the sky!" Long-time comet observer John Bortle, a veteran of the two great comets of 1957 (Arend–Roland and Mrkos), Ikeya–Seki, Bennett, and West, rated Hyakutake as the most spectacular that he had ever seen. His opinion on this did not change following his observations of Hale–Bopp the following year, and he even went so far as to say that whereas in past ages Hale–Bopp would certainly have been regarded as a brilliant comet, the likes of Hyakutake would have been deemed something supernatural and altogether terrifying. This thought is also seen in the following description by Jay Reynolds Freeman, whose impression of the comet on the early morning of March 26 says more than any prosaic scientific description could hope to convey:

> The comet was magnificent, a display that seemed to dazzle our dark-adapted eyes.... I recalled a legend about comets from bygone days, that they are great dragons, bringers of wisdom and knowledge, breathing fire and flame as they speed among the stars, strewing smoke and sparks far across the trembling heavens.

What the ancients saw in the sky usually suggests that they used some of the hallucinogenic substances we regard as modern. Yet this time I could see what they meant: a dragon indeed! To stare at the coma was to gaze into the maw of the beast, the central condensation its very gullet. The straight, narrow beam of the inner tail blazed with the lambent, incandescent blue of a Bunsen flame as the dragon expelled its mighty breath at full force. Farther along the tail the streaming gout of fire lost intensity and coherence as it widened and dissipated in fading swirls of translucent smoke and pale luminosity, through the equatorial Gegenschein and beyond, across half the sky.

Remarkably, this magnificent display was caused by an icy body just 1–3 km across! That, at least, is the conclusion drawn from an experiment in bouncing radar signals off the nucleus as it sped past Earth on March 24–25. Compared with the nucleus of Halley, this is very small, and the fact that it gave rise to a coma intrinsically as bright is truly remarkable. Clearly, this is an extremely active comet that does not depend on a few "hot spots" on its surface. Unlike Halley and most other comets that have been studied in this respect, Hyakutake spread its activity over most, maybe all, of its surface.

Yet, there were apparently two regions on its nucleus where activity was even more vigorous than elsewhere. As the comet passed Earth, large puffs of material were observed being ejected sunward like clockwork every 6.23 h in addition to smaller puffs having the same regularity. Apparently, there were two super active regions – one larger and perhaps more vigorous than the other – that erupted into action each time they came into sunlight. The time between two large and two small eruptions therefore marked the rotation period of the solid nucleus, i.e., the comet's day was just 6.23 h long!

After passing inside Earth's orbit, the intrinsic brightness of the comet fell a little, and its rate of brightening on approach to the Sun slowed. This meant that as it moved away from us, the intrinsic brightening caused by its approach to the Sun failed to compensate for its increasing distance from Earth, and its apparent brightness dropped sharply to about second magnitude in early April. The tail remained very long, however, and between April 8 and 16, lengths of 60, 70, even 80 degrees continued to be reported by observers under optimal conditions. These estimates imply true lengths of at least 80 million kilometers (50 million miles). The comet was also becoming dustier than it had been in March, although it would never be a truly dust-rich object in the manner of most great comets.

Around mid-April, the activity of the nucleus suddenly calmed down. This change was evident, both in the spectrum of the comet as monitored by professional astronomers and in its level of brightness as judged by amateurs and the general public. From then until perihelion, the increase in intrinsic brightness was so slow that some people (mindful, no doubt, of the radar results mentioned earlier) wondered if the nucleus might be shriveling up before their very eyes and whether this once-great comet would disappear completely at perihelion!

The final Earth-bound observation before perihelion was made by John Bortle on April 27, when the comet was just 12 degrees from the Sun and around magnitude 2.5–3. It was, however, picked up and followed through perihelion by the SOHO spacecraft, when it may have shone at first magnitude, and despite some fears to the contrary, re-emerged into the southern hemisphere skies on May 9. That morning, Gordon Garradd of New South Wales, Australia, became the first person to see it following perihelion. It then shone at about the same brightness as Bortle's final sighting pre-perihelion.

This author's own observations of the comet began again on May 14, when I conservatively estimated it as magnitude 3.8 under poor conditions. As the comet emerged from morning twilight during the latter half of May, it became a spectacular binocular object, with a very intense yellow false nucleus, a clearly defined parabolic envelope, and an intense tail visible with the naked eye for up to 3.5 degrees at times. Looking out over a dark ocean horizon, the comet was an easy naked-eye sight, though certainly no longer qualifying as great. Nevertheless, in comparison with the head, the surface brightness of the tail was actually greater than it had been in March, the big difference being that instead of shining at 0–1 magnitude, by late May the coma barely made magnitude 4!

As the comet moved away from the Sun, the very slow brightening noted just prior to perihelion was mirrored by a very slow fading, at least initially. On May 29, the magnitude was estimated as 4.6, with an intense pseudo-nucleus and an impressive tail of 2–3 degrees still visible with the naked eye. The comet remained a naked-eye object throughout most of June, still 5.8 on the 19th as estimated with the unaided eye. Though remaining impressive in 10×50 binoculars, the intensity of the 2.5 degree tail was clearly beginning to wane, and on the 24th the magnitude 6.3 comet had reverted to a more diffuse, globular appearance, without a star-like false nucleus and sporting a faint tail of just one degree. As it receded further from the Sun the rate of fading increased, and the last observation appears to have been on October 24, when the small and condensed object was estimated as magnitude 16.8.

As well as being remembered for its spectacular performance, this comet has become associated with two important scientific discoveries. On March 26–27, it became the first to be observed at an X-ray wavelength – via the German ROSAT satellite – and was found to be emitting this radiation at intensities at least 100 times greater than even the most optimistic predictions. Unlike the comet's visible light, the X-rays were not coming from the central condensation (which was not even visible in the images) but from a crescent-shaped region on the sunward side of the coma. Most of this radiation probably originated with the interactions between solar wind particles and cometary material. Since the Hyakutake discovery, X-rays have been observed in several other comets. Significantly, an observation of C/1999 S4 (LINEAR) in 2000 using the Chandra satellite indicated that the X-rays being emitted from that object were predominantly the products of collisions between nitrogen and oxygen ions in the solar wind and the neutral hydrogen atoms of the comet's hydrogen coma.

The second major scientific discovery was the identification of ethane and methane, the first time that either of these gases had been identified in a comet.

The abundance of the two gases was found to be roughly equal, indicating that Hyakutake formed in a region no warmer than 20 degrees above absolute zero. Presumably it formed well away from the Sun, possibly within a denser than average region of the proto-solar nebula or, just possibly, in another interstellar cloud altogether.

Wherever it may have originated, Hyakutake was not making its first sweep through the inner Solar System in 1996. Like Comet West, however, its orbit has been lengthened by the gravitational boost given it on its latest trip past the Sun's planets, though the effect has not been as extreme as that experienced by the former comet. Although we cannot be too precise about periods of many thousands of years, it seems that an original period of around 15,000 has been increased to at least 72,000, courtesy of its 1996 encounter with the inner planetary system.

Comet C/1995 O1 (Hale–Bopp)

With the exception of Halley, the one comet that could be named by the largest number of people on this planet today would almost certainly be Hale–Bopp. Thanks in part to its prolonged visibility, great brightness, and favorable placement in the sky for the greater part of the world's population, this comet acquired an almost legendary aura seldom seen these days. The comet was discovered on July 23, 1995, by long-time comet observer Alan Hale in New Mexico and independently by Thomas Bopp in Arizona on the same night.

Although Hale had spent many hours sweeping for comets without success in previous years; by 1995 he had essentially abandoned systematic searches and was concentrating more on observing known objects. On the night of the 23rd he planned to observe two comets that he had been monitoring, and after studying the first of these, decided to spend some time browsing through a few deep-sky objects while he waited for the second to rise high enough for observation. One of the deep-sky objects he was looking at was the globular cluster M70, and to his surprise he found that it had gained a companion! A small condensed object of around 11th magnitude appeared alongside the globular in the eyepiece of his telescope. After watching this interloper for about an hour and checking out all possible known objects in the field, Hale noticed the telltale sign of a slight shift in position relative to surrounding stars. A new comet had been found! (We still do not know whether he observed the second comet on his schedule for that night, but we suspect that he did not!).

Bopp's discovery was in some respects even more fortuitous. Bopp had no interest in comets and had never seen one; even Halley's passed him by in 1986! Indeed, he did not even own a telescope, but on the night of the 23rd he was with a group of friends near Stanfield, AZ sharing telescopes for observations of star clusters and galaxies. One of the objects on the list was M70, and it was Bopp who had first call viewing it. So, the first comet he ever saw became his own, shared with Hale in a double title that would soon become known throughout the world!

This author's first observation of the comet came the following day. At a magnification of 71 in a 25-cm (10-in.) reflector, the comet appeared at 3 min of arc across and had a very condensed and compact look about it. Particularly notable was that when a Swan Band filter was used, it became dramatically fainter. This is diagnostic of a very dusty comet and, together with its general appearance, suggested a distant and intrinsically bright object. I estimated the magnitude as 10.7 on that evening. Just four evenings later, July 28, I managed to find it using a pair of 25×100 binoculars at an estimated magnitude of 10.5. At the risk of jumping ahead of the story, let me just mention here that my final binocular sighting of the comet, using the same instrument, did not come until June 15, 1999!

For several weeks following discovery, the orbit of the comet remained very uncertain. Several possible orbits had been calculated, but they differed widely from one another, especially in the date and distance of perihelion. The first attempts had perihelion distances varying from just within the orbit of Saturn to just outside that of Venus! One thing they did agree on, though. The comet was very distant and apparently very large.

Computation of the orbit was helped by prediscovery images of the comet found by T. Dickinson on photographs from May 29 and, especially, by an image dating back to April 27, 1993, found by Rob McNaught of Siding Spring Observatory in Australia. McNaught had noted the comet at the time, but without follow-up observations it had become lost. Interestingly, even then it showed a distinct coma although a very distant 13 AU from the Sun. An attempt by Rob to find it on a plate taken on September 1, 1991, turned up nothing definite, indicating a relatively steep brightening had taken place throughout the first half of the decade of the 1990s. Further prediscovery images also came to light subsequently, but it was the 1993 one that proved most important. It was not quite at the position expected, had the comet been following a nearly parabolic orbit. In fact, from this observation and later ones confirming it, the comet's orbit was found to be elliptical with a period of some 4,200 years. Perihelion was predicted for March 30, 1997, at a distance of 0.99 AU from the Sun. The comet would pass nearest to Earth on March 22 at a rather distant 1.31 AU.

The prediscovery images also disproved another suggestion raised in some quarters following the first indications of the comet's great distance. Not unreasonably, the possibility was suggested that it may have been experiencing a large brightness flare at the time of discovery and that its true intrinsic brightness might be nowhere near as great as first thought. There were some good reasons for thinking this. Apart from the improbably high brightness itself, images of the inner coma revealed a distinct spiral jet-like formation similar to that seen in comet 29P/Schwassmann–Wachmann during its frequent outbursts. This latter object follows an almost circular orbit between

Jupiter and Saturn and has long been noted for its frequent outbursts of up to 5 magnitudes. A similar feature seen in Hale–Bopp might imply that it too was experiencing an outburst of several magnitudes at discovery, and even a comparison between two early photographs appeared to show a contracting and fading of the coma as might be expected if the comet was beginning to wane.

Fortunately, this suggestion proved to be false. The difference in the photographs turned out to be caused by a disparity in the conditions under which they were taken, not by a change in the behavior of the comet itself. Moreover, as more and more observations, including other prediscovery ones, came in, it became clear that the discovery brightness of the comet was not simply a passing phase. Rare though it might be, this was truly one of the intrinsically brightest comets ever seen! To put things into perspective, the comet was still 7.1 AU from the Sun at discovery, and yet was easily visible in small telescopes at 10.5–11.0 magnitude. At a comparable distance in 1984 (on its way to perihelion in 1986), Halley's Comet was about 11,000 times dimmer!

That is not to say that Hale–Bopp did not experience outbursts. Several were noted when it was far from the Sun, both on its way toward perihelion and again on its way out. Some of these increased its brightness modestly, but the effect of others simply intensified the central condensation while leaving the total brightness relatively unaffected.

Some idea of what these were like may be gained from a mistake this author made on August 22. On previous nights, the comet glowed at around magnitude 10, with a small but pronounced central core of approximately 13th magnitude. But on the 22nd I wrote in my notes that I was unable to make a brightness estimate because the comet was positioned over a bright star. The next evening, however, I noted that the star had moved together with the comet! What I had mistaken for a background star was in fact the central condensation in outburst. Yet, the total brightness of the comet had only gone from 10.2 on August 17 to 9.9, six nights later. By the 28th the central condensation had again declined to a small core of about magnitude 13, but the total brightness of the comet was estimated to have declined by just 0.1 magnitude to 10.0.

Brightening remained slow during the rest of 1995 as Hale–Bopp gradually drifted toward evening twilight. The final visual observations for the year appear to have been by Terry Lovejoy on November 23 and Alan Hale himself on November 23 and 24.

The next observation of this comet was a very encouraging one. On February 2, 1996, Terry Lovejoy picked it up low in the dawn using a 25-cm (10-in.) reflector and estimated its magnitude as 8.8, quite a bit brighter than predicted for that date. It continued to brighten nicely during coming months, being estimated by myself as 7.9 in 25×100 binoculars on March 27 and as bright as 7.3 with an 8-min coma in hand held 10×50 field glasses on April 1. Considering the distance of the comet, the April 1 estimate was equivalent to an absolute (H_{10}) magnitude of –3. This is equal intrinsically to the very brightest comets ever seen. Indeed, only two comets have been estimated as bright as this, and neither came as close to either Earth or the Sun as Hale–Bopp was destined to venture.

Some observers were reporting tail lengths of up to half a degree during the third week of April, and on the 28th of that month Terry Lovejoy, using 10×50 binoculars, found the comet as bright as magnitude 6.9, with a coma of 15 min of arc in diameter, i.e., half as wide as the full Moon. This translates to a real diameter of around 2,800,000 km (1,750,000 miles), or over twice the diameter of the Sun! At that time, the comet was very close to an eighth magnitude star and the combined light of both objects was just enough for Terry to glimpse a faint spot with the naked eye. He did not class this as a genuine naked-eye sighting of the comet as he felt that the neither the star nor the comet alone would have quite made the grade. Yet, it must be said that magnitude 6.9 objects have been glimpsed with naked eyes, and, as a matter of fact, my final naked-eye glimpse of Hale–Bopp occurred at a time when I was making it as faint as magnitude 7.3 in binoculars. Terry's sky may not have been quite as dark as the one I was blessed with on that night, but I am inclined to wonder whether Terry would still have glimpsed the comet had no star been present!

In any case, moonlight quickly put an end to seeing the comet naked eye on the following nights, though not for long. The first undisputed naked-eye sighting was by Steve O'Meara (who actually estimated it a little fainter, at 7.2) on May 18, followed by Terry Lovejoy on May 20. By June, with the comet still more than 4 AU from the Sun and 3 AU from Earth, naked-eye observations were already numerous.

However, just as the comet was coming into wider naked-eye accessibility, it experienced a sudden slowing in its rate of brightening. Although its tail continued to slowly take form and its total brightness crept up ever so steadily during the rest of 1996 as it approached Earth and the Sun, there was some real concern that the comet might turn out to be, as one astronomer phrased it, a "Hale-flop"! True, the comet remained one of the intrinsically brightest on record; yet it was becoming clear that the very bright absolute magnitudes evident earlier in the year were not being maintained. Initial forecasts of a brightness maximum equal to Jupiter became decreasingly likely as 1996 progressed, and, although optimists continued to hope for a peak brighter than magnitude zero, other experts were not convinced that the comet would ever be brighter than first magnitude or thereabouts.

According to my own observations, the comet was visible in a tiny 2.5×25 opera glass on May 21 at magnitude 6.6 and a faint (magnitude 5.8) naked-eye object under imperfect skies on June 18. Interestingly, on June 7 (when it was shining at an estimated magnitude 6.1) it could be seen in the same 10×50 binocular field as the fainter periodic Comet Kopff. The two comets actually passed just 3 degrees from each other on June 11. Needless to say, they were not at all close in real space. The intrinsically much dimmer Kopff was simply in the foreground of the still-remote Hale–Bopp.

Having reached 5.8 in June, it is surprising that I was still estimating it as faint as 5.0 as late as October 8, although by that date the tail had become far more impressive in binoculars. After October, northern hemisphere observations indicated that the rate of brightening picked up somewhat as the comet headed toward another conjunction on January 3, 1997. At the time of conjunction, it was still located some 28 degrees from the Sun and north of it, so it was possible for observers in far northern regions to keep it under observation. From Sweden on January 1, Timmo Karhula estimated it as third magnitude and, on the same day, B. H. Granslow in Norway gave it a similar value. Also observing from Norway on that same morning, O. Skilbrei caught it with the naked eye.

In early January – January 2, actually – the Earth passed through the orbital plane of the comet. On such occasions, the dust tail is seen edge on, and any plume of large particles that might accompany a comet may show up as an anti-tail. Although the latter is more often seen after perihelion, on this occasion the comet must have already shed a considerable quantity of coarse dust particles, and an anti-tail 2 degrees long was photographed on January 11 by Bob Yen from the Mojave Desert, CA. As we will see, thanks to the long period of Hale–Bopp's visibility, this anti-tail disappeared and reappeared several times as we passed in and out of the plane of the comet's orbit.

By the end of the month of January, the comet had evolved into an interesting telescopic object showing an exceedingly complex system of jets and spiral patterns within its coma and a very strongly curving dust tail. Magnitude estimates placed it as bright as magnitude 2.5–2.7 on January 31.

With the arrival of February, Hale–Bopp was already being well observed in northern hemisphere's morning skies. Around February 8 and for a short time thereafter, it experienced a rather sudden spurt in the rate of its brightness increase. This later settled down to about the same rate as that between October and February, though stepped up to a slightly brighter level. Two tails were showing distinctly on February 13, a straight plasma tail in addition to the strongly curved plume of dust.

Already in February, the comet was hurling out prodigious amounts of material. According to calculations by Lowell Observatory's David G. Schleicher, at mid-month, when still some 1.2 AU out from the Sun, it was releasing dust at the rate of 400 [metric] tons and water at a rate from 60 to 90 tons (or 18,000–27,000 gallons) *each second!* This was equivalent to the evaporation, *each second*, of a cube of ice measuring 4 m (13 ft) along the side. Putting these results into perspective, at that time Hale–Bopp was releasing 200 times more dust and 20 times more water than Halley and 100 times more dust and 50 times more water than Hyakutake at the same distance from the Sun!

The comet's brightness reached a magnitude of 1.2 on the morning of February 17, according to Jan Vesely in the Czech Republic, who managed to hold the central condensation in view through the 140-magnification eyepiece of a 20-cm (8-in.) reflector until nearly half an hour *after* sunrise. The false nucleus (which was all that remained visible after sunrise) was estimated to be about second magnitude. This became the first of several daylight sightings of the comet during the latter half of February. The trick was to observe it in the pre-dawn sky and, with the advance of twilight, focus on the false nucleus with as much magnification as possible and try to hold it in view for as long as possible after sunrise. It is interesting to note that the comet did not fade below its mid-February levels until the end of May, so it should have been possible to make daylight sightings using moderate-sized telescopes and appropriately high magnification for at least 3 months. Except for an intensive night and day observing program at Mt. Bigelow Observatory from March 19–23, it seems there were no further daytime observations after the comet became a predominately evening object in March, despite its even greater brilliance then. Certainly, it is harder to pick something up before sunset than to follow it through sunrise, but a telescope equipped with setting circles should have enabled experienced observers to find it.

The final day of February saw the comet standing at magnitude 0.5, according to Charles Morris, who noted that this was "significantly brighter" than it had been just 2 days previously. He also noted that "a prominent fountain coming off the nucleus and its resulting dust fan are easily visible in binoculars." Two tails were also visible, but it seems that how much of these could be seen depended critically on local conditions. Thus, while Morris traced the plasma tail for 20 degrees and the dust tail for 10 degrees with the naked eye, John Bortle only recorded 4.3 degrees of plasma tail and 2.2 degrees of dust through 10 × 50 binoculars at the same time. Despite the big difference in recorded lengths, it is interesting that both these very experienced observers roughly agreed on the *ratio* of the tail dimensions.

The first evening observation appears to have been made by Robert Victor on February 23, when he spotted the comet with a pair of 10 × 50 binoculars, very low over the horizon in bright twilight. Nothing was apparent to the naked eye, and the sky was too bright for him to pick out a tail. One month later, the best views were in the evenings and soon thereafter morning visibility ceased altogether.

The comet became a truly spectacular object in early March. According to John Bortle, the plasma tail had increased to 16.5 degrees and the dust tail to 9 degrees by early in the month. Beginning March 6, patterns of striae were observed in the dust tail, reminiscent of those seen in comets such as Donati, Mrkos (1957), Ikeya–Seki, and West. From Central Asia on March 9, observers of a total solar eclipse were afforded the added treat of seeing Hale–Bopp some 46 degrees away from the blacked out Sun.

The comet reached its brightest peak – or perhaps "plateau" would be a better description – between March 26 and April 11. With all bright comets, there is inevitably a scatter in brightness estimates. Nevertheless, once the few reports that are clearly either too bright or too faint are eliminated, the majority of estimates converge between about –0.5 and 0.0. The optimists who predicted early in the apparition that the comet would become brighter than 0 magnitude were right – if only just!

Of great interest during the latter days of March was the amazing amount of detail seen within the coma. Remarkably, these features were visible in telescopes as small as the off-the-shelf 6-cm (2.4-in.) refractors that have served to kick start the careers of so many young amateur astronomers. The most striking features were the series of expanding shells graphically and poetically described by one observer as looking like expanding "ripples in a pond of light." In larger telescopes, a jet or bar could be seen connecting the innermost of these shells, with the false nucleus giving it the appearance, as John Bortle described his observation through a 41-cm (16-in.) reflector at 114 magnification, of a barred spiral galaxy. The bar was the clue to what was happening. These shells were not shock fronts, as some had thought, but sprays of material jetting from active areas on the comet's surface. As the nucleus rotated on its axis, these areas would be periodically exposed to the Sun's heat and burst into activity. First, a jet would erupt outward from the nucleus, but as the latter turned on its axis, this would bend and trail as it tended to wrap itself around the nucleus. This is what Bortle saw as a bar and spiral pattern. Finally, as the active zone passed out of sunlight and shut down, the spiral became detached and expanded outward as a shell. By that time, though, a new active area was already coming into sunlight and a fresh jet starting to emerge. From measurements of the expanding diameters of the different shells, it was determined that they were expanding outward through the coma at approximately 300 m per second. Constantly spiraling jets with almost catherine-wheel patterns probably originated from active zones near the rotational pole of the nucleus (in the comet's "arctic region," so to speak) that were bathed in perpetual sunlight while that pole remained oriented toward the Sun.

From the observations of these jets, a rotation period of about 11.47 h was determined for the nucleus, which was also calculated to be at least 30 (and possibly 50) km (19–31 miles) in diameter. The complex nature of some of the spirals also suggested that the nucleus was wobbling quite wildly on its axis.

As part of an effort to observe this comet as thoroughly as possible, rockets were launched from White Sands on March 25 to image it in ultraviolet light. At this wavelength, the comet's dimensions were truly staggering. Incredibly, the diameter of the neutral hydrogen coma matched Earth's distance from the Sun!

Not only was the hydrogen coma huge, but the central 10 million kilometers (6.3 million miles) or thereabouts were dense enough to block the Sun's ultraviolet light and cast a shadow behind the comet! Observations by the SWAN instrument on board the SOHO spacecraft discovered this ultraviolet shadow cast for more than 150 million kilometers (over 93 million miles) across the hydrogen haze permeating the Solar System. Although not visible in ordinary light, it was the largest shadow ever observed in the Solar System.

A hypothetical being with eyes sensitive only to ultraviolet light would have seen a total eclipse of the Sun by the comet from a location within the cone of this shadow!

This observation was not a mere curiosity for astronomers. It enabled a relatively accurate determination of the amount of water (the parent molecule of the hydrogen) that was being released by the comet – a whopping 300 metric tons per second!

Throughout April, the comet remained a spectacular sight for northern observers. Tail lengths were typically estimated in the 10–20 degree range, largely depending upon the clarity of skies and distance from major sources of light pollution. Maximum lengths of about 20 degrees for the plasma tail were recorded near the time of perihelion passage. Translated into real terms, this is equivalent to around 1 AU or the distance from the Earth to the Sun. Experienced observers under clear and dark skies sometimes gave estimates of 30 or even 40 degrees for the dust tail, and there was one measurement of possibly 60 degrees. April 7 saw the comet looking strikingly reminiscent of the famous painting (probably by Mary Evans) *Donati's Comet over the Conciergerie in Paris on October 5, 1858*. The curving dust tail appeared 18 degrees long, and the 10-degree plasma tail was split into two clearly divided streamers, giving the comet a striking "Donati" appearance. Only the absence of Arcturus near the head broke the illusion of seeing back in time to Paris of 1858! The comet's head was estimated as –0.2 at the time.

Observations from La Palma Observatory between April 16 and 22 led to the discovery of a third type of tail in addition to the familiar plasma and dust varieties – a tail of neutral sodium atoms. Such a tail was not totally unexpected. Sodium had been noted in the heads of many comets in the past, especially those passing well within Earth's orbit, and the first indication of a significant tailward extension of sodium emission was noted by H. F. Newell as long ago as 1910, for the Great Comet C/1910 A1. More recently, sodium emission was traced for 7 degrees along the tail of Comet Mrkos in 1957. But the Hale–Bopp analysis provided the first thorough study and full confirmation of this phenomenon. The sodium tail was not, however, visually distinguishable.

The comet became visible from the southern hemisphere again in late April, and on the 28th was estimated a little fainter at magnitude 0.7. From my own estimates, I judged it to be 0.6 on May 5, 0.8 on May 17, 1.2 on the 24th, and 1.3 on May 27. On that last date, the comet was a spectacular sight in 25 × 100 binoculars, with the main tail having a distinct hook-shaped appearance, rather like a shepherd's crook, curving back sharply from the nucleus. A second very faint ray of a tail formed at an angle of about 50 degrees with the first, and between them there could be seen a broad and faint fan-shaped glow. The whole tail system appeared as a very broad appendage oriented almost parallel with the western horizon.

Fig. 5.10 Comet Hale–Bopp, August 13, 1996 (courtesy, Terry Lovejoy)

Fig. 5.11 Comet Hale–Bopp, April 3, 1997, showing both plasma and dust tails (courtesy, Terry Lovejoy)

The comet continued to fade slowly, from 1.6 on June 3 to 2.0 on June 14. My final evening observation for 1997 was on June 25, when I spotted the comet in 25 × 100 binoculars very low on the horizon and probably about magnitude 2.5, although that is more of a guess than a proper estimate.

Fig. 5.12 Comet Hale–Bopp, May 1, 1997 (courtesy, Case Rijsdijk, SAAO)

My next view came on July 9, when the comet was low in the dawn twilight and estimated at a magnitude of 2.9. As it emerged higher during the following mornings, it presented a very unusual aspect. The comet was upside down! Normally, when a comet emerges into the twilight, the head is near the horizon, and the tail extends upward. There are, of course, very good reasons for this. Bright comets appearing above the rising or setting Sun are almost inevitably close to it in space, and their tails, directed away from it in the normal manner, necessarily point more or less upward. But Hale–Bopp was well beyond the Sun, and we were viewing it from a position close to its orbital plane. Any normal comet would be invisible with the naked eye under these circumstances and would probably show little or no tail even through a telescope. But Hale–Bopp was no normal comet, and its appearance in July 1997 made that abundantly clear. In binoculars and to a lesser degree with the naked eye, the tail appeared to emerge from the very condensed head like a fountain of light and then curve backwards towards the horizon, in the apparent direction of the Sun. Never before had I heard of a comet in the twilight with the tail pointing downwards toward the point of sunrise! The length of tail was about 3.3 degrees as seen in binoculars on July 16, with 1 degree visible to the naked eye. At that time, the total magnitude of the comet was estimated at 3.1. From a darker location on July 31, however, the comet appeared as magnitude 3.0 with 7 degrees of tail visible sans optical aid.

By August 13, the comet had faded to magnitude 4. Around that time astronauts aboard the space shuttle *Discovery* observed it both with the naked eye and through a 7-in. ultraviolet telescope attached to one of the shuttle's windows.

The comet was clearly visible with the naked eye during a total lunar eclipse early in the morning hours of September 17 (16th Universal Time) when I estimated it as magnitude 4.6. I found it 4.7 on the 26th when a broad fan of tail about half a degree long could be seen through 25 × 100 binoculars.

For a time between the end of August and the beginning of November, the comet briefly revisited the northern hemisphere. Suitably, the first person to pick it up was its co-discoverer Alan Hale, who saw it through strong twilight on August 31. A number of other observations were made during that period, including two sightings on October 6 and 7 from as far north as Alaska! The last in this series of second-run northern observations was by H. Dahle, who spied it from Hawaii on November 1 through a pair of 9 × 63 binoculars.

Fading continued for the rest of the year, with estimates of 5.0 on October 3, when a delicately curved 2 degree tail was seen in 25 × 100 binoculars, to 5.8 according to Dahle's November 1 observation mentioned in the previous paragraph. According to my estimates, it had faded to 6.9 on November 25, but brightened again to 6.1 on December 1, as estimated through 2.5 × 25 opera glasses. This was one of a number of weak outbursts experienced by the comet on its way out from the Sun.

As the end of the year approached, Earth once again drew close to the comet's orbital plane, and throughout December the effects of this could be seen as the broad, faint fan of remnant dust tail slowly closed, narrowed, and intensified. In addition, the anti-tail made a comeback. Already on December 3, a tail of 1.2 degrees and an anti-tail of 0.8 degrees were visible in 25 × 100 binoculars. At that time, I estimated the comet's brightness again at 6.1 and faintly detected it with the naked eye. On December 9, Terry Lovejoy also saw the comet with the naked eye, judging it to be magnitude 6.8, and I again managed to glimpse it without any optical aid on December 15, when I gave a rough estimate of 6.4 in the opera glass. The tail continued to intensify, and on December 30, I described the comet as a "striking sight" in 25 × 100 binoculars, with a straight and remarkably intense tail sweeping right across the 2.6 degree field of view and the anti-tail visible for about 20 min of arc. The scene was repeated on January 5, 1998, just one day after Earth crossed the comet's orbital plane. The comet was then estimated at magnitude 7.2. The following night, Gordon Garradd at Tamworth, New South Wales, recorded 1.5 degrees of anti-tail in a 25-cm reflector and traced the main tail for about 4.5 degrees in 10 × 50 binoculars.

The tail maintained its length through the first half of January, with some 3 degrees visible in 10 × 50 binoculars on January 16 and at least 5 degrees in the same instrument on the following night. Although estimated at magnitude

7.3 in the binoculars, by using averted vision I actually caught a clear glimpse of the comet with the naked eye on January 16, and with less certainty the following night. As far as I am aware, this was the last time that the comet was seen with the naked eye. This means that Hale–Bopp was a naked-eye object for nearly 21 months if we class Terry Lovejoy's April 1996 sighting as being of the comet itself and not just the combined comet/star conjunction (see above), or 20 months if we count from O'Meara's May observation. Either way, it was over twice as long as the previous 9-month record set by the Great Comet of 1811 and is not likely to be equaled for a long, long time.

At the time of these observations, the anti-tail remained visible, though not as prominently as earlier in the month. The main tail remained relatively obvious until early April, by which time the total magnitude had faded to near 8.

Even as late as January 5, 1999, the comet was estimated as 9.7 in 15 × 80 binoculars. An outburst had occurred in mid-December 1998 when Gordon Garradd noted a rise of 3 magnitudes in the brightness of the central condensation, and this central outburst evidently left the comet unusually bright for several weeks. A conical tail up to 7 min of arc long made it an attractive sight in my 25-cm (10-in.) reflector. The comet's brightness then fell quickly to 11.0 on February 4, after which a slow fading once more set in. Remarkably, I could still glimpse it using 25 × 100 binoculars as late as June 15, 1999, when the magnitude had faded to 11.5.

More than 10 years have gone by since this gigantic comet passed perihelion, yet, remarkably, as these words are written, it is still within the range of large telescopes equipped with CCDs. Who knows when the final observation will be made? Some have even predicted that it will be followed until 2020 or thereabouts!

The long period of visibility allowed the orbit to be determined with a precision seldom possible for long-period comets. Unlike its predecessors Hyakutake and West, Hale–Bopp's passage through the inner Solar System actually caused its orbit to become smaller and its period shorter. When it entered the inner planetary system on its current visit, it was moving in a relatively elongated ellipse with a period of around 4,200 years, but thanks to the gravitational tug of the planets (principally the massive Jupiter), this has been shortened to just 2,380 years. It is possible that even more drastic changes will take place in the future, and according to some calculations, there is a 15 percent possibility that Hale–Bopp will one day end up in a sungrazing orbit. Unless its intrinsic brightness has faded significantly by that time (if, indeed, "that time" ever comes!), imagine the spectacle it will provide then!

Incidentally, talking of grand spectacles, astronomer Dan Green has pointed out that if the comet had reached perihelion 4 months earlier, it would have passed just 0.11 AU from Earth in early January, about the same minimum distance as Hyakutake passed in March the previous year. Moreover, Hale–Bopp would have been placed equally favorably in the sky as seen from the northern hemisphere at that time, riding high all night long at a magnitude of at least −5. It would have been easily visible with the naked eye in full daylight and cast clear shadows at night. Perhaps it has already put on such a grand display at an earlier return or may yet do so in the distant future. In that case, Hale–Bopp may have been, or may yet be, not just *one of* the greatest of the greats, but *the* greatest comet of all!

Comet C/2006 P1 (NcNaught)

The early years of this new century have witnessed so many comets discovered by professional sky patrols that reports of new ones are now treated as routine! Most of the comets discovered by these patrols are too faint for visual observation at the time of discovery, and the majority remain that way, being either intrinsically too faint or too remote (or both!) to be of any practical interest to the visual observer.

Accordingly, the announcement that Robert McNaught of Siding Spring Observatory in New South Wales had discovered yet another 17th magnitude comet on August 7, 2006, during the course of the Siding Spring Survey (SSS) aroused little excitement. This was Rob's 31st named comet discovery (a few others were not immediately recognized as comets and were given the name "Siding Spring"). Most of his discoveries were found during the SSS or its predecessor, although he did find one in an amateur capacity back in 1987. (By the way, the very week these words were written, his 35th comet discovery was announced!)

A very preliminary orbit suggested a perihelion distance near the orbit of Mars around mid 2007. Not especially exciting!

In the days following the comet's discovery, however, new accurate positional measurements showed that its track through the sky was departing from the preliminary orbit and indicated that the date of perihelion would fall a lot sooner than first thought; *and* at a greatly reduced distance from the Sun. The new orbit yielded a perihelion date around January 11 at just 0.17 AU from the Sun. (This was later refined to January 12, but with the perihelion distance remaining essentially unchanged.) Other things being equal, a small perihelion distance means a bright comet, but in this instance it seemed that those "other things" were far from favorable.

Most disturbing was the apparent faintness of the comet. Both the discovery report and early brightness estimates put it at magnitude 17 or fainter. According to the data to hand, the absolute magnitude of the comet was at best 10. This was not encouraging, and not simply because it implied dim magnitudes in January. Most astronomers saw this as effectively sounding the comet's death knell!

Although it had long been suspected that intrinsically faint non-periodic comets with small perihelia had a strong tendency to fade out completely near the Sun (those of short period are a different matter and do not show this tendency), it was not until John Bortle made a detailed study of these objects in the early 1990s that a definite statistical relationship between a comet's absolute magnitude and perihelion distance was discovered. By omitting comets of short period and those with perihelia lying outside of Earth's orbit, Bortle found that a limiting absolute magnitude (in terms of H_{10}) existed for every perihelion distance, at least out to 1 AU, where his study stopped. This relationship could be represented by the formula,

$$H_{10}\text{Lim} = 7.0 + 6q$$

where $H_{10}\text{Lim}$ is the faintest absolute magnitude at which a comet is likely to survive and q the perihelion distance in AU.

Although there is some indication that this may not always hold for comets with moderately long periods (hundreds to low thousands of years), it has proven to be very reliable for those with very long periods as well as the ones that appear to be making their first journey from the Oort Cloud. In fact, the only "exceptions" here have not been fainter comets surviving but ones brighter than the predicted limiting magnitude that have *not* survived! It seems that the formula is better at predicting which comets will fade out (which, after all, was the purpose of Bortle's study) than which ones will survive.

For a comet like 2006 P1, having a perihelion distance of 0.17 AU, the formula predicts a limiting survival magnitude of

$$7.0 + 6 \times 0.17 = 8.02$$

To use an expression that has become popular among comet enthusiasts, the "Bortle Limit" of a comet with a perihelion of 0.17 is an absolute magnitude of approximately 8. But when the new orbit was published in mid-August, all indications were that the absolute magnitude of P1 was just 10, which was 2 magnitudes *fainter* than the Bortle Limit. This is why most comet observers were pessimistic about its future prospects.

I personally wondered, initially, if the available estimates might be too faint. CCD estimates frequently are conservative, sometimes by as much as 3 magnitudes, and because the only estimates reported by the middle of August had been byproducts of positional measurements, they probably were on the faint side. Astronomers making positional measurements are trying to record the *least* coma, not the most. They need small images to work with, so their magnitude estimates are likely to fall well short of the true total brightness of the comet, unless the coma really is very small and compact. On further thought, though, I came to suspect that this one probably was very compact. The brightness estimates being reported were in good agreement with Rob's discovery observation, and I suspected that this may not have been too far off the mark. Several weeks earlier, Rob had discovered another comet, which he estimated as about 14 in the survey images. This is rather bright for a survey discovery, and I tried for it visually with the 25-cm (10-in.) reflector, wondering if it might be even brighter than the discovery report suggested. Alas, although it was visible in the telescope, it was actually less than a magnitude brighter than Rob had estimated, and I figured that if this were true for his earlier comet, it was probably true for 2006 P1 as well. At best, his estimate was likely to be just 1 magnitude too conservative, still putting the comet below the Bortle Limit.

The first indication that this might not be the whole story came on August 21. That evening, Terry Lovejoy was continuing his regular patrol of the sky with a digital camera (with which, happily, he was later to discover two comets of his own during the first half of 2007!) when he came across a diffuse bluish smudge just a couple of minutes of arc across and of total magnitude around 14. Clearly it was a dim comet, and Terry checked the position and found it to be exactly at the predicted position of C/2006 P1 but around 3 magnitudes brighter!

On learning of Terry's observation, I tried for it visually with the 25-cm (10-in.) reflector but initially met with more frustration than success. Unfortunately, the comet was located against the band of the Milky Way. Fortunately, though, it was moving slowly along one of the dark dust lanes where there were few stars. But again unfortunately, there were *so* few stars visible in its immediate vicinity that fixing its precise location in the eyepiece field was no easy task! Finding the comet's position relative to field stars was quite easy in the wide field of a low-power eyepiece, but the separation of the field stars was such that they lay outside the smaller field of a high-magnification eyepiece and, if the comet was to be seen at all, high power was needed to give sufficient contrast against a dark field. However, going from low to high power meant going to a blank starless field. It was impossible even to get the right focus, let alone zero in on a marginally visible comet!

After one or two unsuccessful attempts, on the evening of August 25 I noted that the comet had shifted into a "mini constellation" of 12–14 magnitude stars small enough to fit into the field of the high-power eyepiece, yet the stars remained sufficiently sparse to avoid interfering with the comet image. It seemed ideal as I peered hopefully into the eyepiece and saw … nothing, at least not straight away. But, just as I was starting to think that I had been too ambitious, there occurred one of those periods of extra clarity beloved by visual observers. Stars that had been marginal came into sharp focus, and those that previously had been too faint to see now popped into view. And, in the middle of the field, there it was! A barely discernible smudge like a tiny fingerprint, my first view of C/2006 P1!

By comparing it with defocused images of nearby stars of known brightness, I estimated it to be magnitude 13.9 and some 0.6 min of arc in diameter. This was quite a bit smaller than Terry's estimate, so I was probably only seeing the brighter central regions of the coma, and the magnitude estimate was likely to be still a little conservative. Nevertheless, even taking Terry's estimate and mine at face value was enough to lift the comet over the Bortle Limit, and for the first time provide some hope of its survival. My August 25 observation was, incidentally, the first visual sighting of this comet.

During September, a number of other people spotted the comet visually, but it remained very diffuse and was far from being an easy object. Because of its diffuse appearance and lack of obvious central condensation as observed visually, some observers were still not convinced that it would survive perihelion. In answer to this skepticism, Rob McNaught pointed out that his CCD observations *did* show a very pronounced condensation and that this appeared to be brightening faster than the total magnitude of the coma. The reason that visual observers were not seeing this was simply because it was too faint at that time, but if its brightening rate continued unabated, the comet should become sharply condensed as it drew closer to the Sun.

October brought some better news, with a few estimates placing the comet between magnitude 11 and 12 in the first half of the month. On the 11th, I even managed to see it faintly through a 10-cm (4-in.) binocular telescope at 25 magnification. It was very diffuse, not unlike the Gegenschein appears with the unaided eye on a clear and very dark night, and was estimated as 6 min of arc in diameter with a total magnitude of 11.1. The estimate of the coma size, considering the comet's distance from Earth at the time, translated to a true diameter close to 679,000 km. This was confirmed on CCD images taken by Michael Jager and G. Rhemann in Austria on that same date. These observers registered an outer coma of 6–7 min of arc surrounding a more intense inner region 2–2.5 min in diameter. They estimated the total CCD magnitude as 12. It is interesting to note that the comet's visibility was enhanced somewhat when a Swan Band filter was used, indicating that the lion's share of the outer coma's light came from the emission lines of the usual cometary gases. This is instructive in view of the later development of this comet into an extremely dusty one and cautions us not to rely too strongly on a comet's early performance as an indicator of its later development.

The comet was observed on October 12 by Jaun Jose Gonzalez in Spain when at an altitude of just 10 degrees. In his 20-cm (8-in.) reflector, Gonzalez estimated the coma as 2 min of arc across and glowing at magnitude 11.7. This observer again saw the comet on November 9 at an altitude of only 5 degrees, estimating it to have brightened to magnitude 9.8. By the 13th it had further increased to 9.3 and to 9.1 on the 16th as again estimated by Gonzalez. On this last night, it was also observed by Gary Kronk and estimated by him at magnitude 9.3.

From mid-November through to mid-December, the comet was very poorly placed, and few, if any, observations were made. Michael Jager located it on images taken on December 16, and Polish amateur Piotr Guzik saw it visually in a 20-cm (8-in.) reflector deep in twilight on Boxing Day as a strongly condensed 4.5 magnitude object just 30 s of arc in diameter. Rob McNaught's prediction that the comet would eventually show a sharp condensation had been spectacularly validated and it was becoming clear that this really was an active comet with the potential to become an impressive object. Using a 25-cm (10-in.) reflector and CCD, K. Kadota in Japan saw the comet on December 30 and 31, estimating the magnitude as 3.8, and the coma diameter as 1.7 min on the first date, and 3.5 and 2.0, respectively, on the second. On the second date, he also saw a tail around 4 s of arc long, an inconspicuous beginning to the magnificent spectacle that would later follow.

From the relatively few observations spanning the early October to late December period, it was evident that the comet was brightening quickly and developing in a way that promised a spectacular display in January. Questions about its ability to survive were no longer being voiced, but there remained a lot of doubt as to how bright the comet would actually become. Few astronomers, in late December, were willing to go on record predicting a really brilliant display, and the majority probably expected the brightening rate to come to an abrupt halt early in the New Year. There were good reasons for this restraint. An upgraded orbit published in December showed that the comet was apparently making its first approach from the Oort Cloud. While there are exceptions, enough of these first-time comets have been underachievers for a note of caution to be sounded.

Fortunately, such restraint was unnecessary. Come New Year's Day, and David Moore of Dublin, Ireland, found the comet in 20 × 80 binoculars just 3 degrees above the horizon, estimating its brightness as magnitude 2.5–3. The following evening Gonzalez, using 25 × 100 binoculars, also saw it at an altitude of just 3 degrees and estimated it as magnitude 2.7. According to this observer, the coma was 1.5 min of arc across and sported a short dust tail one tenth of a degree long. On January 3 Haakon Dahle, observing from Norway, estimated the comet as 1.5 in 9 × 63 binoculars and found that, once located in the binoculars, he could then find it with the naked eye. The following day, he estimated it to be as bright as first magnitude with the naked eye. It was also possible to see the comet in both the evening and morning twilight in early January, and on the morning of the 5th B. Leitner saw it easily with a pair of 8 × 30 binoculars, estimating the brightness as 1.5.

As the second week of January began, the comet was becoming a conspicuous naked-eye object with a bright dust tail low in the northwest after sunset and in the northeast before dawn. Those in far northerly latitudes had the best view, and with the short mid-winter days meaning that many people were at work until after dark, it is not surprising that the first view of the comet for many people came through the windows of city office blocks. In the north of Scandinavia, where the winter Sun remained just below the horizon, the comet was seen all day like a pillar of yellow-orange fire floating above the hidden Sun. On at least one occasion it was mistaken for the contrail of a jet aircraft, until its relatively fixed position in the sky revealed its true nature. Further south in the United States, the comet became easier to see as the second week of January progressed, and once again it attracted wide attention from the general public. There is even one story of a motorist driving home from work who got such a shock upon seeing it that he drove his vehicle up onto a footpath! Fortunately, nobody was hurt and no serious damage done.

Considering how badly placed the comet was at the time (just 15 degrees from the Sun on January 5 and only 12 by January 10), it is a testimony to its brightness and the intensity of its tail that it attracted this much attention from the general public. But its very low altitude possibly made it *more* conspicuous to the public rather than less. Was it Lord Baden Powell who advised climbing a tree if trying to hide, on the premise that most people seldom look up? But to see this comet, nobody needed to look up. Its location very close to the horizon made it eye level, so to speak, for anyone going about his or her normal business!

By the second week of January, the false nucleus had become so brilliant that it was hard to believe that this was the same object as that uncondensed diffuse patch of light visible back in October. From being essentially invisible then, the condensation had become so intense in early January as to be compared by one observer to a piece of burning magnesium!

On January 7, Dennis di Cicco found the comet in broad daylight through a 5-in. refractor equipped with a Go-To mount. On that same day, Terry Lovejoy secured an image of the comet during the daytime. This was the start of a period of daylight visibility extending through to January 16, during which it is probable that 2006 P1 was seen by more people during daylight hours than any previous comet. During its brightest period, from January 13–14, the comet was visible with the naked eye in broad daylight simply by screening the Sun from view behind the corner of a building or even an outstretched hand.

Unlike Hale–Bopp, Comet McNaught did not display (at least visually) a great deal of structure within its head. Three rather inconspicuous semi-circular arcs or shells similar in kind to those seen in Hale–Bopp were noted by Nicolas Bivar (France) on January 9 and by Alfredo Pereira (Portugal) on the 11th, but these never became as conspicuous as their Hale–Bopp counterparts. Also on that latter date, Timo Karhula (Sweden) suspected an elongation of the false nucleus or even a hint that it was splitting. There is further evidence that a minor split did occur around that time. Late on January 10 Richard Keen, in Colorado, noted two condensations of magnitude −2 and −1, separated by some 10–15 s of arc while the comet was high in the daytime sky, but nothing further was reported of this double nucleus, and it seems that, if the secondary condensation was indeed a solid fragment and not merely a knot of material, it was very small and quickly evaporated. Certainly no major Comet West-type disruption occurred in this comet.

The comet's stability around the time of perihelion was fortunate in so far as it provided a great opportunity to look for the presence of brightness enhancement due to forward scattering, uncontaminated by any intrinsic outburst caused by a rupturing nucleus. Fortuitously, Dr. Joseph Marcus was in the latter stages of revising a detailed paper on this subject in which he analyzed data from several previous comets and convincingly showed that the process is an important one for those dusty objects that pass between Earth and the Sun. Marcus had long been interested in the role played by forward scattering in comet brightness, and his paper represents far and away the most detailed study of the subject to that time. But modeling behavior in retrospect is one thing; predicting what will happen in a specific future case is another, and the power of Marcus' study would be greatly enhanced if it could be used to accurately predict the behavior of a comet ahead of time. With a great sense of timing, McNaught was just what the doctor ordered!

Marcus determined that the best forward-scattering geometry would occur on January 14, 2 days after the comet's perihelion, when he predicted a

2.3 magnitude enhancement of brightness due to forward scattering alone. This is where the comet's stability was critical. If it split and flared in the manner of Comet West (which also passed through good forward-scattering geometry), it would be difficult to work out what part of the excess brightness was due to forward scattering and what was a genuine flare. Fortunately, McNaught behaved itself over the critical period, with nothing observed that would indicate unusual activity within the nucleus.

Using the latest brightness estimates from early January until just before perihelion, Marcus determined that the comet was brightening at a rate approximating $n = 4$. Comets do not necessarily fade at the same rate at which they brighten, so Marcus allowed for two sets of possibilities, one assuming a symmetrical light curve (fading and brightening at the same rate) and one assuming a slower rate of fading. Following from this, he calculated four sets of magnitude predictions, one each for $n = 4$ and $n = 3$ in the absence of forward scattering and another pair for the same parameters, but including his calculations of the forward-scattering effect. Without going into too much detail here, a summary of the magnitude predictions immediately following perihelion gave

Date (Universal Time)	$n = 3$ (no FS)	$n = 4$ (no FS)	$n = 3$ (+ FS)	$n = 4$ (+ FS)
Jan. 12.0	−3.4	−3.4	−4.1	−4.1
Jan. 12.5	−3.5	−3.5	−4.6	−4.6
Jan. 13.0	−3.5	−3.5	−5.1	−5.1
Jan. 13.5	−3.5	−3.5	−5.5	−5.5
Jan. 14.0	−3.4	−3.4	−5.7	−5.7
Jan. 14.5	−3.3	−3.3	−5.7	−5.6

The first thing that strikes us with this table is that if forward scattering is operative, the peak brightness falls not at perihelion but two days later, irrespective of the value of the brightness parameters. So the first test was to see whether the comet peaked near perihelion (in which case, forward scattering would not be present) or 2 days later, as the model forecast. The second test would be to monitor the actual magnitudes attained by the comet.

The comet did indeed peak at the time Marcus predicted, not at perihelion passage itself. Moreover, the excess brightness was, within an acceptable margin of error, just as he had calculated, with most observers estimating maximum magnitude to be between −5.5 and −6 on January 14. For a bright object in the daytime sky, the estimates were actually in remarkable agreement and indicate magnitudes of around −3 on January 11, close to −4 on the 12th, −5 or a little brighter on the 13th, between −5.5 and −6 on the 14th, and about −4.5 the following day. The predictions of the $n = 3$ and $n = 4$ light curves were too close to distinguish over that range, but the comet's behavior later in January was closer to the $n = 4$ prediction, indicating that the brightness development of the comet was quite symmetrical either side of perihelion. This well-behaved symmetry is further indication that the brightness peak on the 14th was indeed due to forward scatter and not an intrinsic flaring of the comet.

The comet passed through the fields of the Sun-monitoring spacecraft STEREO and SOHO beginning January 11 and 12, respectively. The SOHO image was so saturated that the tail looked solid, but STEREO captured fine detail that gave the tail a remarkable feather-like appearance. This was the first indication of the patterns of striae that were to become so spectacular just over a week later.

In addition to the curving dust tail, STEREO images also recorded what at first sight looked like an ion or a plasma tail. This feature was a good deal fainter than the dust tail and appeared straight and narrow but lacked the fine ray structure normally associated with plasma tails. Analysis by Marco Fulle (of the National Institute for Astrophysics in Italy) and colleagues showed that although this may have superficially looked like a plasma tail, it was actually something quite different and, indeed, represented the first evidence for a previously unrecognized type of tail – not an *ion* tail but an *iron* tail! A tail of neutral iron atoms were swept out of the comet by the pressure of sunlight. This iron is thought to have been supplied by the evaporation of fluffy grains of triolite.

Following perihelion, the comet passed from northern to southern skies, and from January 14 southern hemisphere observers began noting it in the southwestern sky following sunset. On the evening of January 16, the bright yellow dust tail could be seen for about 2 degrees in very bright twilight, and on the 18th the curving scimitar of the dust tail extended for at least 10 degrees and was so intense that it remained visible long after the head of the comet had set. By then, the tail was clear enough for the striae to be noticeable, but nothing prepared observers for what was to come the following night, when the tail assumed a form unprecedented in modern times.

The day of January 19 was, for my location, relatively clear but with some streaks of smoke haze visible in the west by sunset. As the sky darkened, the comet became a commanding spectacle, with a good 15 degrees of bright tail sweeping upwards and merging with what I assumed to be one of the layers of smoke haze that I had noticed earlier. This "haze layer" appeared parallel to the horizon and seemed like a dimly glowing band of vapor extending for another 15 degrees from the point where the tail met it. It looked like an extension of the tail, but as that would mean that the tail made a 90-degree turn and arched northward, I assumed that this was not possible, and the haze layer was simply that – a layer of atmospheric haze illuminated by the dying embers of twilight or simply reflecting back suburban light pollution. Still, it did look like the tail!

I think it was with some relief when the following day I saw posted on the Internet a photograph of the comet from the previous night clearly showing the now-famous arch-shaped tail. What I saw had indeed been the tail, something that, quite frankly, I did not think possible. The following night I paid it more attention! Sweeping along the tail with 10 × 50 binoculars, the striae were nothing short of amazing. Near the region where the tail turned toward the north, bright bands were interspaced with relatively dark gaps that some observers even referred to as "dark striae." The (bright!) striae were obvious with the naked eye and looked like the beams of an aurora frozen in time. There was an especially bright and extensive one going up through the constellation of Grus. This feature would actually have made an impressive comet tail by itself!

According to Marco Fulle's analysis, around this time the dust tail reached about 1 AU in length and, at the broadest part, in the vicinity of 65 million kilometers (just over 40 million miles) across. It is noteworthy that the longest striae also extended across the full 65 million kilometers of tail width!

One of the best views of the monstrous dust tail was by Terry Lovejoy, who viewed the comet from a location about 200 km (125 miles) inland from Brisbane, Queensland. Observing from relatively low latitude, he could trace the tail further north than more southerly observers, even though the comet head was closer to the horizon than for observers at higher southern latitudes. As he saw it, the tail arched northward from the comet head, across the western sky, and down to the horizon in the northwest. So long was the tail that even in mid-northern hemisphere latitudes, where the comet itself remained out of sight well below the horizon, the striae of the extended tail were observed protruding like long fingers of light from above the western horizon. The scene was compared by John Bortle and others to the famous drawings by de Cheseaux et al. of the Great Comet of 1744 (see Chapter 4).

Fig. 5.13 Comet 2007 P1 McNaught over the Sydney Harbor Bridge, January 19, 2007. © 2007 David Austen, used with permission

Fig. 5.14 Comet McNaught reflects in the waters of Lake Leslie, Queensland, on the evening of January 20, 2007 (courtesy, Terry Lovejoy)

Fig. 5.15 The magnificent dust tail of Comet McNaught, January 21, 2007 (courtesy, Terry Lovejoy)

From Hawaii on January 20, Steve O'Meara was amazed to see a curtain of striae projecting from beneath the western horizon along an arc of some 35 degrees. The tallest "were about 20 degrees in extent." O'Meara remarked that the spectacle "was like seeing 8 to 10 comets with tails all at once!" or like "dim searchlights in the sky." He commented that "not even [Comet West] had that kind of sweeping grandeur – and I'm just seeing the tip of the tail!"

As in the northern hemisphere prior to perihelion, the comet attracted wide notice from the general public after it slipped south of the Sun. It was so bright and fiery looking that on one occasion it was even mistaken for an aircraft on fire.

Like C/1910 A1, which it resembled in so many respects, this comet was extremely dusty and likewise displayed strong sodium emission while near the Sun. An extension of this emission into the tail – reaching to 100,000 km (62,500 miles) on January 29 – apparently marked the presence of a sodium tail, albeit not as extensive as the one detected in Hale–Bopp a decade earlier.

Many observers were mystified by the complete absence of a plasma tail when the comet first became visible in tolerably dark skies during January. Unlike most comets that show strong dust tails, the companion straight blue ion tail was strangely missing. Gas must be present for the dust to be expelled from the nucleus, and it was suggested that the dissociation of gas molecules was so rapid that the plasma tail simply did not have a chance to form. A weak plasma tail did, however, develop later in January and showed up in color photographs taken by Rob McNaught as a blue feature contrasting with the yellow of the far brighter dust tail; however, it never became a conspicuous feature of this comet.

Although not apparent visually, images of the central condensation secured by the European Southern Observatory Multi Mode Telescope on January 29 and for several days thereafter captured a catherine wheel of three jets spiraling out from a point nucleus within a region approximately 1 min of arc in diameter. These appear to have been gaseous, although a dust feature was also visible as a bright fan that emerged on the sunward side before sweeping back into the comet's dust tail. At that time, the normal cometary emission lines were visible in the spectrum in addition to the strong solar continuum reflected from dust. The lines of neutral sodium were, however, by far the strongest features in the emission spectrum at the end of January.

By January's end, the comet's brightness had declined to a still respectable second magnitude, and tail lengths of 30 degrees continued to be reported with the naked eye and up to 40 degrees photographically. From dark skies outside the rural town of Cowra in the New South Wales wheat belt, I could still see the comet clearly with the unaided eye at third magnitude early in the second week of February. At least 30 degrees of tail were clearly visible with the naked eye, but it may have extended far beyond this. In fact, On February 8, it still seemed to be faintly visible up to 80 or even 90 degrees from the head. Many would dispute this (some have let their skepticism be known in no uncertain terms), but that was the impression I had. Maybe someday, someone will come forward with a deep all-sky photograph that will prove or disprove this once and for all!

In addition to the main tail, an anti-tail also became visible in February. As seen in a 25 × 100 binocular telescope on February 14, this feature extended about 12 min of arc opposite the main tail, i.e., in a more or less sunward direction. This was not simply a projection effect, but actually involved heavy particles lying between the comet and the Sun. At times it looked almost like a tube of mist, reminiscent of similar description of the sunward feature accompanying the Great Comet of 1882, albeit on a far smaller scale.

The comet faded steadily through February and March. On March 9, the magnitude was estimated at 6.4, although from the dark sky site, some 10 degrees of tail remained dimly visible without optical aid. March 11 saw it at 6.7 in the binocular telescope, with the tail still plainly visible and the anti-tail remaining apparent as a "faint sheath of light" enveloping the coma and extending in a direction opposite that of the main tail. Even as late as March 14, when the comet's head was estimated as magnitude 7.1 and invisible to the naked eye even from a dark site, some 2 degrees of tail could be seen very faintly with the naked eye. This is something that happens from time to time with dusty comets. The total light of the tail can remain high enough for naked-eye visibility even after the comet itself has become an exclusively binocular object. I also experienced this with Ikeya–Seki in early December 1965, and Dutch observers noted that the tail of C/1980 Y1 (Bradfield) was visible with the naked eye in twilight as a faint sheath of light, even though the comet itself could not be seen without binoculars. We wonder, in passing, how many final observations of historic comets were actually observations of the tail alone, and if any of these have affected later calculations of absolute magnitudes of these objects.

By the middle of 2007, as these words are being written, C/2006 P1 is nothing more than a diffuse tenth magnitude patch of gossamer on the edge of the Milky Way in the constellation of Musca the Fly. Presumably, it will be traced for a while yet with the aid of larger and larger telescopes. As it passed through the inner Solar System, the comet was first accelerated to hyperbolic velocity relative to the Sun, but forward calculations show that this will change as it pulls away from the central planetary system and that its orbit will eventually revert to an ellipse, with aphelion falling short of Oort Cloud distances. Eventually the comet will return, though not for 90,000 years or thereabouts!

How did Comet McNaught compare as a spectacle with other great comets of recent times? As far as comparison with the two great comets of the 1990s is concerned, here are the words of Terry Lovejoy, who had the enviable fortune of seeing all three under near ideal conditions around the times of their maximum displays.

His verdict? McNaught "left Hyakutake and Hale–Bopp for dead!"

We might also recall that Bortle judged Hyakutake to have been the best that he had seen, and he witnessed every great comet since 1957. Unfortunately, though, this veteran observer did not see McNaught at its best, so he was never in a position to truly compare the two.

Other comet observers dispute the relative glories of McNaught and Ikeya–Seki, but whatever the final decision, the Great Daylight Comet of 2007 surely rates as one of the most magnificent of recent centuries.

With Comet McNaught, we reach the most recent participant in the parade of history's greatest recorded comets, a parade that began when the philosopher Aristotle was a boy of just 12 years. We have no idea when the next great comet will appear, and we have no idea whether it will be one of the greatest of this class or not. But whether it is another Hale–Bopp or McNaught or simply a spectacular comet in its own right, it will undoubtedly leave in its wake another generation of comet enthusiasts with their curiosity whetted by this spectacular and unusual sight in the night skies.

Chapter 6
Kamikaze Comets: The Kreutz Sungrazers

The year 1843 saw the appearance of a remarkable comet, rated by many as one of the most spectacular of the nineteenth century. We will have more to say about this comet later, but for the moment let's simply note its amazingly small perihelion distance, just 0.0055 AU, well within the Sun's corona! (It is not for nothing that comets such as this one have in more recent times been likened to the Kamikaze suicide pilots of World War II!).

Fast forward now to February 1880, when southern hemisphere observers were again startled by the sudden appearance of a great comet in the south-western evening skies. Although the new visitor became neither as bright nor as spectacular as the one 37 years earlier, its appearance as a small and compact coma at the base of a very long tail of remarkably uniform intensity was very reminiscent of the earlier object. But the orbit provided the real surprise. It was essentially identical with that of the earlier comet!

It seemed most obvious to simply assume that the two comets were consecutive appearances of a single periodic object. After all, by the late 1800s there were already a number of known precedents!

Yet, there were problems with this straightforward explanation. For one thing, no similar comet had appeared around 1806 (the Great Comet seen the following year had a very different orbit) and, secondly, the orbit of the 1843 comet suggested a period of several hundreds of years, while the less well-determined orbit derived for the comet of 1880 did not clearly depart from a parabola. If it really was elliptical – as was very probable – its period had to be a lot longer than 37 years.

The first difficulty might be overcome if the comet's period was shortened dramatically at each return through the resistance of the Sun's corona. If that were true, the apparition previous to 1843 might have been a lot further back than 1806. Indeed, the orbit could be made to fit the paths of a comet in 1668, and it also seemed possible that the Great Daylight Comet of 1106 was a still earlier apparition. Some astronomers even thought it possible that the orbital period may vary drastically from one return to another, some being longer and some shorter. Variations in the period of a comet are normal, but the extent of those proposed in this instance bordered on the absurd. The "average" period

D. Seargent, *The Greatest Comets in History*,
DOI 10.1007/978-0-387-09513-4_6, © Springer Science+Business Media, LLC 2009

was supposed to be just 7 years, even though 37 years had elapsed between the most recent returns and even longer intervals occurred throughout history.

The hypothesis of a contracting period may have briefly been given a boost in 1882 when yet another comet was discovered moving in a similar orbit. However, because this one came to be so well observed over a longer arc than either of its immediate predecessors a more accurate orbit was computed, and this left no doubt that the true period was in the hundreds of years. This was not another instance of a short-period comet being seen at different returns, but of truly individual comets all moving in very similar orbits, albeit differing significantly in the dates of their perihelia. It was, in fact, the first recognized example of a rare phenomenon, a comet group.

The orbits of some of these comets were studied by Heinrich Carl Friedrich Kreutz, who concluded that the orbital elements were so similar that the individual comets most probably arose from the disruption of a single progenitor as it passed very near the Sun many hundreds of years before. A similar suggestion had earlier been made by Martin Hoek with respect to comets traveling in similar orbits. The idea of comets breaking apart and the pieces becoming individual objects in their own right had become respectable since the splitting of the short-period Comet Biela in 1846 and its return as two comets in 1852. (As it turned out, however, most of these earlier suspected associations were simply statistical clusterings and not true groups.) Hoek knew nothing of the 1843/1880/1882 group, however, as only the first of the trio had appeared during his lifetime, and it was Kreutz who has been honored by having his name associated with this particular comet group.

A further "sungrazer" (as these comets came to be known) appeared in 1887, followed by a very faint and poorly observed one in 1945. In the middle years of last century, the Kreutz comets were pretty much relegated to history, until the surprise appearance in 1963 of Comet Pereyra and, especially, the magnificent Ikeya–Seki of 1965, the most brilliant comet of the twentieth century and a spectacle in the same league as the historic objects of 1843 and 1882. These were quickly followed by Comet White–Ortiz–Bolelli in mid-year of 1970. Like Pereyra, however, this latter object was poorly placed and did not provide a great display. Nevertheless, had the sungrazers of 1963 and 1970 appeared at more favorable times of year, they would have equaled Ikeya–Seki itself!

These events, in particular the 1965 comet, renewed interest in the group, and in 1967 Dr. B. G. Marsden undertook a fresh study of the sungrazing comets in the light of recent observations. Marsden published a second and more comprehensive paper in 1989, and more recently, Z. Sekanina and Paul Chodas have undertaken continuing research into the group's evolution.

Fresh impetus was given to the study of these comets following the launch of satellite-based coronagraphs and their unexpected discovery of prodigious numbers of tiny members of the group evaporating near the sun. The first indication of this came when the SOLWIND coronagraph on board US Navy satellite P78-1 imaged a small comet apparently striking the Sun on August 30, 1979. Because the information was not released until 1981, nothing was known about it at the time, but during the years following, both SOLWIND and the Solar Max satellites found several similar sungrazing comets, and, since 1996, the SOHO coronagraphs have found well over 1,000; the nuclei of most have been estimated as being less than 10 m (33 ft) in diameter!

Marsden's initial study was undertaken after the sungrazers of the 1960s but prior to that of 1970. He, therefore, had four good orbits to work with, those of the comets of 1843 and especially 1882, 1963, and 1965. Tracing these backward in time, he found that the positions of the 1882 and 1965 comets coincided to within the bounds of acceptable error at the time of their previous perihelion passage, indicating that the split into separate objects occurred then. From the computed periods of both comets, this perihelion passage presumably took place during the very early years of the twelfth century. As we shall see below, both these comets were observed to split, and the fragments were well-enough observed for separation velocities to be derived. Significantly, Marsden found that the separation velocity of the two comets themselves at their previous return was of a similar magnitude, apparently confirming his suspicion that these were the major fragments of a comet that broke apart near the previous perihelion passage some time in the early 1100s. The Great Daylight Comet of 1106 seemed an obvious choice as the parent.

The orbital histories of the 1843 and 1963 comets were also investigated in this study, but Marsden found that they could not be brought together in the same way at the previous perihelion passage, although they were certainly closer then. They may have split from a single parent at the passage prior to the previous one, early in the first millennium. The period of Pereyra suggested a previous apparition around the year 1100. If the 1106 comet was *not* the 1882/Ikeya–Seki combo, it would be a good candidate for the previous return of Pereyra, but a comet seen in China in 1075 was also a possibility. Although its orbit was not as well determined, the 1880 comet moved so similarly to that of 1843 that a common origin at *their* previous return seemed more probable.

Marsden noted that, although the sungrazers had appeared in "clusters," of greater importance was the apparent existence of two subgroups, one sharing orbits very similar to that of the 1843 comet and the other having orbits closer to those of 1882 and 1965. He termed these Subgroups I and II, respectively, and argued that each probably originated with the disruption of two comets that were themselves the main sub-nuclei of a giant comet that split some time between five and 20 revolutions earlier. The differences in the orbits of the subgroups are subtle. Perihelia of the first subgroup are a little smaller than the second, and the values for the node and argument of perihelion are different, but it is clear that they are still very closely related. He noted, however, that the "clusters" of sungrazers appearing in different centuries contained members of both sub-groups and were probably just statistical groupings.

His study was updated in a second paper published 22 years later, following the discovery of the 1970 sungrazer and the first of the small objects found by SOLWIND and SOLAR MAX. Essentially, his analysis remained unchanged, except that he included the 1970 comet as a single member of the sub-subgroup

IIa. Its orbit clearly shared more in common with the mean orbit of Subgroup II than with Subgroup I, but more as an outlier rather than a core member of the subgroup itself.

Marsden's updated study also reduced the time since the grandparent of the sungrazers was disrupted. The forces, both thermal and tidal, experienced by these comets when close to perihelion are tremendous, and they surely cannot survive many such encounters with the Sun. The shortening of the past history of the group is therefore an advantage for any analysis, and Marsden's upgrade suggested that the original sungrazer may have been the comet seen by Aristotle around the year 372 B.C. This might have been the comet that split into the parents of the subgroups. The 1882/1965 split remained unchanged (probably 1106), but the new analysis suggested that another (unidentified) Subgroup I comet around that same time split into two major fragments, one which achieved a period of around 900 years and returned in 1963 as Comet Pereyra, while the other retained the parent's period of approximately 400 years and returned unobserved in the late 1400s, the most probable date being late October 1487. This hypothetical comet – to which Marsden gave the name "Combo" – split again at that perihelion passage, the fragments returning as the comets of 1843, 1880, and (most probably) the small comet known as "Tewfik" and observed only during the total solar eclipse of May 17, 1882. Presumably the other sungrazers separated at one or other of these former perihelion passages.

With the advent of SOHO, the water has become muddier. What may have looked like relatively well-defined subgroups became blurred in a wide range of variations on the Kreutz orbit theme. Another feature of the SOHO comets is the tendency for these to cluster, sometimes more than one being simultaneously visible in the coronagraph field! While groupings are to be expected in any random distribution, studies by Zdenek Sekanina and Paul Chodas indicate that many of these clusters are real and result from disruptions of single objects. However, these disruptions cannot have taken place at a previous perihelion passage. Compare, for instance, the 83-year separation between the 1882 comet and Ikeya–Seki, as fragments of a disruption the previous time around, and the separations of only hours for some of the SOHO cluster members. Moreover, the SOHO comets are so tiny that they evaporate before perihelion, and even their immediate parents are unlikely to have been capable of safely completing a single sungrazing passage. Apparently, the SOHO clusters are the end products of a series of fragmentation events happening long after the previous perihelion passage. Presumably, the immediate parents of the cluster members broke away from surviving split nuclei sometime following the previous perihelion and these in turn fragmented further at a still later date. In fact, it was probably not the parents of the clusters that separated relatively early after the previous perihelion. It may have been their grandparents or great-grandparents! Fragmentation, according to Sekanina and Chodas, is a progressive process that occurs at any distance from the Sun and at any point in a comet's orbit.

Continuing analysis by Sekanina and Chodas of the Kreutz group's evolution, taking into account this "cascading" feature of nucleus disruption, has thrown up some surprises. For instance, according to their 2004 paper, the orbit of Ikeya–Seki could be better accounted for if the comet split away from a combined 1882/1965 nucleus some 18 days after the previous perihelion rather than at perihelion itself. In another surprise, Pereyra appeared to be dynamically more closely related to 1882 and Ikeya–Seki than to the comet of 1843, despite the greater similarity between its orbit to that of the latter-mentioned one. If this conclusion is correct, the validity of the subgroups is called into question, and the tendency of the large sungrazers to cluster at certain dates may be more important than the earlier Marsden analysis suggested. Moreover, if Pereyra is not as closely associated with 1843, the necessity to suppose an unobserved "Combo comet" in the late 1400s is no longer required. The indication of a common origin for the comets of 1843 and 1880 is, however, strengthened in the 2004 paper, and the 1887 comet is added to this dynamical cluster as well.

As this chapter was being written, yet another Sekanina/Chodas paper was published making even more radical departures from the simple scenario of the early Marsden research. This new approach suggests that the 1106 comet may *not* have been the parent of the 1882 and 1965 objects after all! Instead, it may have been an object with a "Subgroup I" type orbit that tidally split into a train of progressively disintegrating fragments immersed in a cloud of debris. These fragments, when far from the Sun, kept breaking up non-tidally to form the SOHO sungrazers as well as some of the major Kreutz objects of more recent times. The 1106 comet and the true parent of the 1882/1965 pair are, according to this new paper, thought to have parted company in the non-tidal splitting of a "grandparent" comet on it way to a fifth century perihelion. The authors hypothetically link the 1106 comet with a "sungrazer suspect" of February 423 or, alternatively, with the spectacular but rather poorly documented comet seen in February 467. They find that if the latter link is correct, sungrazer clusters in the 1600s become possible. Moreover, this link suggests that the progenitor of the entire group may have been seen (albeit very poorly documented) as the comet of 214 B.C. (not 372 B.C., as has been widely suspected). Neither the comet of 467 nor that of 214 B.C. feature in most lists of suspected sungrazers, and the preserved accounts of both objects yield little information (see Chapter 3 of this book for the meager details available on the fifth-century comet).

An interesting conclusion of this new research is the suggestion that we have not seen the last of the major sungrazers and that further clusters may keep occurring until 2120. The next cluster is probably only decades away on this scenario. In fact, its early members may show up just a few years from now!

Almost certainly, the full story is far from being told, and we can expect further refinements in the years to come. The best advice is probably to "watch this space" and, if the latest Sekanina and Chodas paper is correct in its prediction, it might prove profitable to keep a watch on the sky as well!

How Many Kreutz Comets Are There?

Excluding the 1,000+ mini-comets recorded by SOHO coronagraphs, how many Kreutz comets have been recorded through the ages?

The definite members of the group are the comets C/1843 D1, C/1880 C1, C/1882 R1, C/1887 B1, C/1945 X1 (du Toit), C/1963 R1 (Pereyra), C/1965 S1 (Ikeya–Seki), and C/1970 K1 (White–Ortiz–Bolelli). In the past, some doubts were raised about the membership of 1887 B1, but it now seems safe to include it as a bona fide member.

In addition to these, we should probably include X/1882 K1, a small comet seen only during the total solar eclipse observed from Sawhaj in Egypt on May 17, 1882. Although no orbit could be computed and the comet was not seen again, Marsden pointed out in his 1967 paper that the position of this comet was very close to that expected for a comet moving in the 1843 orbit just 5 h before perihelion. This comet, by the way, is always referred to as "Tewfik," named in honor of the Khedive for his hospitality extended to the eclipse expedition.

Other, earlier possible Kreutz comets have been pointed out by a number of people, principally by I. Hasegawa as part of his broader study of possible past apparitions of recent comets. The principal possibilities are the comets of 214 B.C., 372 B.C., A.D. 191, 252, 302, 423, 467, 852, 943, 1034, 1106, 1232, 1381, 1666, 1668, 1689, 1695, and 1702. Marsden also cited the comet of 1075 as a possible former apparition of Pereyra and the "blazing starre seen near unto the Sonne" reported on Palm Sunday 1077 has also been included as a possible sungrazer, although A. Pingre thought that Venus was a preferable explanation for this event. Similarly, an object seen near the Sun from Broughty Ferry (Scotland) at 11 a.m., December 21, 1882, has likewise been suggested as another possible sungrazer – or Venus! This object was said to have had a "milky appearance" and to have been crescent-shaped when observed "through the glass," but no trace of tail was reported.

In 2002, R. Strom of the University of Amsterdam drew attention to Chinese records of mysterious "Sun stars," or bright star-like objects apparently seen within the disc of the Sun. He argues that, although it is difficult to understand what these could have been if they were truly *within* the solar disc, the wording of the records leaves open the possibility that they were simply very close to the Sun, in which case they might have been bright comets near perihelion. At least some of them may have been members of the Kreutz group, aqccording to Strom. "Sun stars" are listed for the years 15 A.D. (sometime between March 10 and April 7), 1539 (between July 15 and August 13), 1564 (between August 7 and September 4), 1625 (September 2), 1630 (August 5), 1643 (between June 16 and July 15), 1644 (between March 9 and April 6), 1647 (July 28), 1648 (in the summer), 1665 (August 27), 1774, 1792 (between April 24 and May 20), 1839 (August 14) and 1865 (July 18). None of these can be associated with a comet, although a "star" was said to have been seen in the daytime from London at the time of the birth of the second son of Charles I on May 29, 1648. This may have been the "summer" object recorded in China.

An interesting aspect of the "Sun stars" is the number that appeared during the May/August period. At that time, as Strom points out, Kreutz comets approach and recede in the direction of the Sun, and from the end of May until late August they can only be seen, if at all, very close to the Sun in daylight. Earlier in May or from the end of August, they might be seen in twilight (cf. Comets Pereyra and White–Ortiz–Bolelli) but only at low elevations and exclusively from southern latitudes.

Also of interest is the clustering that took place during the seventeenth century. If the "Sun stars" really were comets, and if a number were members of the Kreutz group, there must have been a lot more activity from this group in the 1600s than is generally thought.

Yet, there are no definitive ways of proving whether or not these "Sun stars" really were comets. Were they some form of meteorological phenomenon? The most obvious candidates – parhelia or Sun dogs – appear to be ruled out, however, as these were well known to the Chinese by the seventeenth century.

Is it possible that they were naked-eye white-light solar flares? Occasional entries in Chinese records of solar events have been suspected of this interpretation. For example, on December 9, 1638, a large sunspot is recorded followed by the note that "black and blue and white vapors" also appeared within the Sun. On the other hand, solar flares appear together with sunspots, but there is only a single instance of a "Sun star" and sunspot being mentioned together.

Interpreting these observations depends on just how we read "in the Sun." Many of the records of both "Sun stars" and sunspots say that they appeared "in the Sun," and if we understand the sense of the expression in the same way in both sets of accounts, it would seem that the "Sun stars" – like sunspots – appeared within the solar disc. This all but rules out the comet explanation. Yet, as we hinted earlier, some of the reports can be translated as "at the Sun's side," and one record (for 1839) specifically reads "in the Sun a star appears, its position is at the Sun's side." Strom notes that 30% of the records give the "Sun star's" position as "at the sun's side." But then, what exactly does this mean? Does it imply (as Strom believes) that the "star" is at the side (i.e., close to but clear of) the solar disc, or does it mean that the "star" is actually within the disc, but closer to the limb than to the center? Both interpretations are possible.

One more point to ponder. The 1600s, when the lion's share of these "Sun star" observations were made, was also the century of the so called Maunder Minimum, a period of low-solar activity that extended from 1645 until 1715. "Sun stars" were reported near the beginning of this period, but not then until the late 1700's. Also, the period of the Dalton Minimum (1795–1825) appears to have been "Sun star" free. Whether this has any significance is not clear.

It would be helpful if just one of the daylight comets seen close to the Sun from countries other than China during the last millennium could be equated with a "Sun star," but such is not the case. The daylight comets of 1106, 1402, 1577, 1744, and 1882 were not recorded by the Chinese in daytime hours, while that of 1843 is clearly designated as a "broom star" when observed in the daytime sky. This comet, by the way, appeared just 4 years after a "Sun star," and a further "Sun star" was recorded 22 years later, yet there seems to have been no comparison made between these objects and the 1843 daylight comet. Moreover, it seems odd (though, of course, far from impossible) that the above-mentioned daylight comets should have escaped detection in China while an impressive list of otherwise unknown objects were spied so close to the Sun as to be described as actually located within it!

It may also be worth mentioning that the daytime sungrazers of 1843, 1882, and 1965 also displayed clear tails in daylight. Had earlier objects of similar appearance been recorded by Chinese astronomers, they would very likely have been designated as daylight "broom stars," (as indeed, one appearing in 302 was!), thereby dispelling all doubt as to their nature!

We have already looked at the possible sungrazers of 372 B.C., 191, 252, 467, and 1106 in Chapter 3. The first and last of these certainly ranked among the "greatest of the greats," and from what little information we have about the other three, they seem to have been up among the ranks as well, whether or not they were bona fide members of the Kreutz group.

We will now take a quick look at the other possible early sungrazers, plus some objects of the late 1600s and early 1700s that are widely thought to have been Kreutz comets, before turning to the definite members of the group and, in particular, the three premier members, 1843 D1, 1882 R1, and 1965 S1.

Possible Early Sungrazers

214 B.C.

Although not formerly suspected as being a sungrazer and not found in Hasegawa's list, this comet was cited in the 2007 Sekanina/Chodas paper as a candidate for the progenitor of the entire Kreutz group. The year of its appearance may agree with their model, but the records of the object are too vague to confirm or deny an association with the group. The sole account of the comet is given in the *Shih chi* and simply states that a "bright star appeared in the west" during that year. No date is given, and there has even been debate as to whether the account refers to a comet or nova, although the commentary on the *Shih chi* does call it a "broom star," apparently confirming it as a comet.

A.D. 133

A Chinese source documents a comet that year on February 8, having a tail about 75 degrees long and 3 degrees wide. If it was a Kreutz comet, perihelion would have occurred on January 20, give or take 3 days.

302

A Chinese text records that a "broom star" was seen in the daytime sometime during the period May 14 to June 11. Nothing is said as to how long it remained visible or how widely it was seen. We cannot say whether it was seen only once by one person or whether it was widely apparent, but the lack of any nighttime observations suggests that it was *only* seen in daylight and, probably, only briefly. The reason for thinking that this may have been a Kreutz object (besides its obvious brilliance) is that a Kreutz comet appearing mid-year would approach and recede from behind the Sun and, if visible at all, would be seen only in daylight. The Chinese account is repeated in a Korean chronicle, but no extra information is provided.

423

Chinese chronicles note a bright comet in February and indicate a tail up to 45 degrees at its best. The positions given suggested to Hasegawa that this may have been a Kreutz comet with perihelion between February 3 and 9, but he also noted that its location in the sky made it a candidate for an earlier apparition of the "sunskirting," though definitely non-Kreutz, comet discovered by P. M. Ryves in 1931, assuming perihelion on January 29, ± 5 days. There are, however, wide discrepancies in the estimates of the period of Ryves' Comet with the most recent computations (by Marsden and E. Everhart in 1983), indicating a far longer period than the earlier ones known at the time of Hasegawa's paper. According to the Marsden/Everhart orbit, any earlier perihelion passage would have occurred far back in prehistory, precluding any chance of identity with the comet of 423.

852

Chinese chronicles report a "star" with a tail of, perhaps, 75 degrees length appearing in mid-March of that year. Little further detail is given, but Hasegawa suggests that the location of the comet is consistent with a Kreutz orbit having perihelion between March 4 and 10.

943

Chinese texts describe a "broom star" with a 15-degree tail appearing on November 5 in the constellation of Virgo. Hasegawa suggests that the observations could fit a Kreutz orbit with perihelion between October 27 and 29.

1034

This comet was seen by the Chinese on September 20 and was described as having a tail of 10.5 degrees long and less than 1 degree wide. Brief records of the comet are also found in Japanese and European records and, according to Hasegawa, the positions are consistent with a Kreutz orbit with perihelion between September 3 and 13.

1232

This "broom star" was first seen in China on October 16 and described as having a tail 15 degrees long and "bent like an elephant's tusk." Interestingly, the similarity to an elephant's tusk recalls a description by an Indian observer of the tail of the Great Comet of 1882. The tail of the earlier object had extended to 30 degrees by October 26, but the comet could not be seen in moonlight on October 30. This indicates that it had become rather faint, yet after the Moon left the sky, it was again visible, and a tail length as great as 60 degrees was recorded for November 10. It seems to have remained visible until the middle of December. A Kreutz orbit gives perihelion between October 8 and 18.

X/1381 V1

This comet appeared on the morning of November 7 with a tail 15 degrees long and remained visible until December 11. If it was a member of the Kreutz group, perihelion would have fallen between October 29 and November 2.

None of these comets was observed sufficiently well for even an approximate orbit to be computed. That the rough positions and times were apparently consistent with Kreutz group membership may or may not be significant.

Between the years 1668 and 1702, however, there appeared four comets whose membership in the group has been strongly suspected, even though non-Kreutz orbits have been satisfactorily computed for three of them (the fourth was insufficiently observed for any orbit to be determined). We refer to the comets of 1668, 1689, 1695, and 1702.

C/1668 E1

This comet appeared in the evening skies of March 3 with an intense tail over 30 degrees long. Initially, the tail was so bright that it cast a clear reflection in the sea, but by March 6 it had become, in the words of an observer in Brazil, "so thin that the eye could easily see the stars that were behind it." By March 30, the comet had disappeared.

In the course of his discussion on "Sun stars," Strom noted that this comet was possibly seen in daylight in China. He based this on an ambiguity in Chinese records between "evening" and "daylight," pointing out that in the Chinese records of the 1843 comet, where one record says that it was seen in daylight, the other says "sunset" or "evening." While granting this, the Chinese records of C/1668 E1 do not commence until March 7, by which time Western records clearly indicate that it had faded significantly and would not have been visible in daylight.

The "official" orbit of this comet (ironically, computed by none other than H. Kreutz!) is *not* a Kreutz group orbit, even though it is "sunskirting." Perihelion date is February 28, 2008, at 0.07 AU from the Sun. Nevertheless, the path of the comet is almost as well represented by a true Kreutz orbit with perihelion on March 1.4 at 0.005 AU. This orbit is very similar to that of the 1843 comet.

Incidentally, there is an observation of a bright comet made by Robert Knox during the time he was held captive in Ceylon. He recorded this as February 1666, but the comet he described was so similar to that of 1668 that an error in the year was long suspected. Nevertheless, a Korean record also has come to light of a comet seen early in 1666 and it is possible that these accounts refer to yet another Kreutz object!

C/1689 X1

This comet was first observed on November 24 by Simon van der Stel at the Cape of Good Hope, just before sunrise. It then passed very close to the Sun and emerged again from bright twilight on December 7, when it was seen from the French East-India Company ship *Jeux*. Tail lengths in the 45–47 degree range were recorded around mid-December, reaching 60 degrees on the 21st when the comet was located close to Centaurus. Nevertheless, by that time it had already grown faint and disappeared altogether during the first days of January.

Once again, the "official" orbit is not that of the Kreutz group (perihelion being given as November 30.66 at 0.06 AU), but the path of the comet may also be represented by a Kreutz orbit if perihelion occurred on December 2 and the other orbital elements were similar to those of the 1882 sungrazer.

C/1695 U1

This comet was discovered on October 28 by P. Jacob, a French Jesuit living in Brazil. He saw it near the eastern horizon 1 h before sunrise, with a westward-pointing tail. As it emerged from the morning twilight, tail lengths of up to 40 degrees were recorded by the end of the first week of November. Never-theless, even as the tail extended, the comet rapidly faded and was last seen on November 18.

The "official" orbit gives the date of perihelion as October 23.77 at 0.042 AU; however, the comet's path can also be represented by a Kreutz orbit similar to that of the 1843 comet, and perihelion at October 23.1.

X/1702 D1

This comet was first reported from ships sailing near the Cape of Good Hope on February 20 and was said to have shown a tail over 42 degrees long. Despite the length of tail, however, it does not appear to have been especially bright. "A cometary blaze, fine and dim" was one telling description. Unfortunately, there were very few sightings of the comet and only three tolerably accurate positions exist. These, however, lie rather close to what would be expected for a comet in the orbit of the 1882 sungrazer, assuming a perihelion date of February 15.1. The comet was last seen on March 1.

Undisputed Sungrazers

Turning now to the "core" members of the Kreutz group, we can immediately name three as qualified for being truly great comets. Surely, the great daylight comets of 1843, 1882, and 1965 were among the greatest of the great comets of history and well deserving of a place here. The other five (1880, 1887, 1945, 1963, and 1970) we will call the "minor" group members, but we must be clear as to what we mean by "minor" in this context. Two of these objects (1963 and 1970) were intrinsically anything but "minor' and had they been more favorably placed, would most probably have rivaled those of 1843 and 1965. The comets of 1880 and 1887 are each referred to as the "great comet" of their respective years, although they were clearly fainter than the other great sungrazers. Only the intrinsically faint and poorly observed 1945 member was truly "minor" in the proper sense of the word. In addition to these, of course, we also have the myriad sungrazing mini-comets observed from space in recent decades, but as none of these has been seen from the ground, they are outside the scope of this book.

The "Minor" Objects

C/1880 C1

Although sometimes referred to as "Gould's Comet," Gould was really only one of many discoverers on February 2, and it seems that an unnamed "gentleman in the northern part of [New South Wales]" was actually the first to sight it on February 1. Charles Todd of Adelaide Observatory described the tail on February 2 as appearing "as a narrow whitish auroral streak ... the upper portion curving somewhat to the south." It was then about 20 degrees long, but by February 5 and 6, observers were reporting lengths of 50 and 60 degrees, even 75 degrees according to David Gill at the Cape of Good Hope. The head, however, was small and inconspicuous, and descriptions indicated a level of brightness no greater than third magnitude at the time when these great tail lengths were being recorded. By February 13, the comet had become difficult to see with the naked eye and after the 20th had completely disappeared.

Fig. 6.1 The Great Comet of 1880 being observed by amateur astronomers, February 14, 1880. Courtesy, State Library of Victoria

Incidentally, Strom also mentioned this comet as having possibly been seen in daylight by the Chinese, for the same reasons already noted with respect to the 1668 object. However, once again, this seems improbable, as the Chinese record of the comet describes it as a "broom star" seen in the southwest, probably around February 7–9. It is unlikely to have been visible from anywhere in China before then, but from observations throughout the southern hemisphere, it is clear that the comet was already growing rather faint by the end of the first week of February and daylight visibility was out of the question. Interestingly, when this comet was thought to have been a return of the 1843

Fig. 6.2 The Great Comet of 1880 from Melbourne Observatory, February 16, 1880. Wood engraving published in *The Illustrated Australian News*. Courtesy, State Library of Victoria

object, the lack of daytime observations were specifically cited as evidence that it had not been as bright in 1880 as at its "previous return"!

Perihelion occurred on *January 28.12* at just 0.0055 AU.

C/1887 B1

Sometimes called "Thome's Comet" and sometimes "the Headless Wonder," this comet was actually discovered by a farmer in South Africa on January 18 and by J. M. Thome at Cordoba the following evening. It was a striking object for several nights, with tail lengths in excess of 40 degrees being measured between January 23 and 25. A curious feature of the comet, however, was the lack of any definite head. Most observers simply recorded the tail as dwindling off to nothing in the field of vision! Although a remarkably interesting appearance, this also made it very difficult to find the proper position of the [nonexistent?] nucleus, which in turn made computation of an orbit very unreliable.

This is the main reason why genuine Kreutz group membership of this comet was at times questioned.

The comet faded quickly, and the last person to see it seems to have been J. Tebbutt at Windsor on January 30.

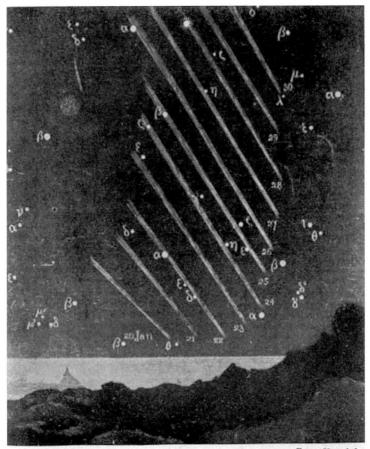

From *Knowledge*

THE SOUTHERN COMET OF JANUARY, 1887

Fig. 6.3 The "Headless Wonder" of 1887, showing the motion of the tail from January 20–30, 1887. From Mary Proctor's *Romance of the Comets* 1926, courtesy, R. Bouma

By studying the available observations, Z. Sekanina computed a Kreutz orbit, similar to that of 1843, having perihelion on *January 1.93* at just 0.0048 AU, the smallest for any comet excepting some of those discovered in SOHO and other satellite date. Because of the peculiar circumstances surrounding this comet, however, the perihelion distance remains somewhat hypothetical.

Sekanina also found that the tail was in reality a single synchrone band released in a burst of particles only hours after perihelion. It seems that this burst marked the total disruption of the comet's small nucleus, accounting for the strange "headless" form that it had assumed by the time of discovery. It was, quite literally, a decapitated comet tail!

C/1945 X1 (du Toit)

With an absolute magnitude of just 10.8, this is the intrinsically faintest member of the Kreutz group thus far observed from the ground. It was discovered photographically by D. du Toit at Bloemfontein in South Africa on December 11, 1945, as a diffuse tail-less object of magnitude 7, and followed for only 4 more days before becoming difficult in twilight. An orbit similar to that of the 1882 sungrazer was computed from the available photographic observations spanning a total of just 5 days. Perihelion came on December 28 at 0.008 AU from the Sun. It is likely that the comet faded out completely before perihelion, as neither the comet nor an 1887-type "headless tail" emerged from the evening twilight in early January 1946.

C/1963 R1 (Pereyra)

This comet was discovered on September 14, 1963, by Zenon M. Pereyra of the Cordoba Observatory in Argentina and estimated as second magnitude at that time. Although it is not impossible that the comet was experiencing a flare at discovery, it is more likely that the reported discovery brightness was overestimated, as it was judged at just sixth magnitude by Alan McClure (Hollywood, California) on the 16th when a 10.5 degree tail was seen and photographed. Through McClure's binoculars, the tail resembled a faint searchlight beam rising out of the dawn twilight.

Fading rapidly as it receded from both the Sun and Earth, the comet was down to magnitude 7 on September 23 (when the false nucleus was gauged at magnitude 13.2) and 7.5 by the month's end. The nuclear condensation was estimated as magnitude 17.2 on November 9, according to Elizabeth Roemer at the US Naval Observatory (Flagstaff, AZ) when a probable secondary condensation was also suspected 0.1 min of arc from the main one. Further photographs were obtained by K. Tomita (Dodaira, Japan) on November 16 and 26 and by Roemer on December 14 and 18. On this last date – the final observation of the comet – the nuclear condensation was estimated as magnitude 18.2.

This comet has an orbit strikingly similar to those of 1843 and 1880, with perihelion on August 24 at just 0.005 AU. The perihelion distance of Pereyra is actually slightly less than those of the 1843 and 1880 comets, making this the smallest well-determined perihelion distance of any known comet. (The smaller ones computed for 1887 and some of the mini-sungrazers of recent decades are less well determined.)

C/1970 K1 (White–Ortiz–Bolelli)

Shortly after sunset on May 18, 1970, Mr. Graeme Lindsey White, university student of Barrack Point, New South Wales (later to become Dr. Graeme White, professional astronomer) was using an old pair of 12 x 50 binoculars to sweep the bright western twilight for possible comets near the Sun when he happened upon a first-magnitude highly condensed object with a stumpy tail. At the time, the comet was just 12 degrees from the Sun. White spotted it again on May 20, this time with the naked eye as well as in binoculars, and by then, the tail had grown to around 10 degrees.

The following evening, an independent discovery was made by Air France pilot Emilio Ortiz, flying east of Madagascar. He saw the comet from the cockpit, estimating its brightness as magnitude 0.5–1 and the tail as 5–8 degrees. Just a few hours later, a second independent discovery was made by Carlos Bolelli, a technician at the Cerro Tololo Interamerican Observatory in Chile. Bolelli did not see the head of the comet (which had already set), but his attention was caught by the long tail projecting from beneath the western horizon.

May 24 was probably the best time to observe the comet, as it was then at sufficient elongation and, shining at an estimated magnitude 4, still bright enough to be easily visible. Tail lengths to around 15 degrees were reported on that date. On May 31, it had faded to magnitude 6. On June 2, some 6 degrees of tail were clearly visible through my 20 x 65 binoculars, and the comet had the appearance of a miniature (actually, a more distant) Ikeya–Seki. By June 6, however, the comet had faded to around magnitude 9, with the head appearing as a diffuse and transparent globule, although up to 1 degree of tail remained visible. Fading quickly and sinking back into evening twilight, the comet was last observed on June 7 by A. Jones (New Zealand) and M. Jones (Queensland) at magnitude 9, just 22 degrees from the Sun.

Although unquestionably a member of the Kreutz group, the orbit of C/1970 K1 does not exactly correspond with either the 1843 or 1882 type, although clearly closer to the latter. Perihelion occurred on May 14 at 0.009 AU, the largest well-determined perihelion distance for any member of this group.

This completes our quick survey of the less brilliant of the undisputed sungrazers observed from the ground. Once again, however, we stress that the last two objects were intrinsically much brighter and potentially far more spectacular than they actually appeared. Had the first arrived at perihelion a couple of months later and the second equally earlier, they would have become great comets; maybe even numbered amongst the greatest of the greats. Indeed, had either arrived at perihelion during late December or January, it may have become more spectacular than any other Kreutz comet!

On the other hand, if C/1963 R1 had arrived just a couple of *weeks* early and C/1970 K1 a couple of weeks late, each would have been so poorly placed that we most probably would never have known of its existence.

The Greatest Sungrazers

C/1843 D1

This magnificent object, sometimes known as the Great March Comet, became one of the most splendid ever seen and is often cited as the finest of the nineteenth century. For instance, in November of 1843, T. Maclear recalled that "[the comet of 1811] was not half so brilliant as the late one." Likewise, after observing Donati's Comet, Ewart remarked that, despite its splendor, "it was, however, nothing to the one I had seen in the year 1843." Furthermore, replying to a statement that the comet of 1882 was "the finest in 200 years," J. Lefroy opined that "[the author of that statement] could not have seen the one in 1843."

Nevertheless, its entry onto the astronomical stage was anything but auspicious, and it seems that the first reports of its presence were made, not in an official publication by a famous astronomer, but anonymously in a New York newspaper! The very first sighting apparently came on the evening of February 5, and a subsequent report dating from February 11 placed the comet "in the vicinity of β Ceti." Nothing more was seen (or at least reported) until February 19, 23, and 26, when a number of vague and scattered sightings of a comet tail in the evening twilight eventually came to light.

Then, on February 27.66 UT, P. Ray at Concepcion, Chile, independently spotted the comet in broad daylight near the limb of the Sun. This marked the first daytime sighting and seems to have been the only occasion before perihelion when the comet was observed in that way. Nevertheless, the observation has always presented a problem; when the orbit was eventually computed, it placed the comet further from the limb of the Sun than Ray had reported. According to Ray, the comet was just 5 min of arc from the limb of the Sun at the time of his observation, whereas the computed orbit suggests it to have been about 1.5 degrees from the Sun's center. This is more realistic for the sighting of an object in daylight, especially one whose presence was not previously known. Although there seems no doubt that Ray really did see the comet, it is difficult to understand how he could have confused a separation of around 3 solar diameters for one of just 1/6 of a solar diameter. Perhaps the "5" was a misprint for "50," still an underestimate, but a more understandable one!

In any case, just 5 h later, the comet passed behind the Sun (although, not surprisingly, this occultation passed unseen!) and remained in eclipse until February 27.91, at which time it was also, coincidentally, passing through its perihelion just 0.0055 AU from the Sun's center. Nobody saw this jewel hanging on the Sun's limb, however, and it seems that nobody saw the comet move away from the Sun immediately following perihelion, then turn and move back toward it (from the perspective of an observer on Earth) before transiting the solar face between February 27.97 and 28.02.

By February 28.4, the comet had withdrawn far enough from the Sun's blazing disk to become remarkably conspicuous in the daylight sky and was independently discovered by numerous folk all over the world. In China, it was recorded as "a large broom star" near the Sun. From Italy, it was seen as "a very beautiful star" with a tail as long as 4–5 degrees. One observer in North America traced the tail for a remarkable 8–10 degrees through a telescope in broad daylight! Estimates of around 3 degrees were common when the comet was just over 4 degrees from the Sun, and both head and tail were said to have been as sharply defined as the Moon on a clear day.

The wide extent of daylight visibility was probably due to the length and intensity of the tail as much as to the brightness of the comet per se. Without doubt, the comet was incredibly brilliant – surely in the –10 range – but the length of tail as seen against the daytime sky must have greatly enhanced its visibility. It is interesting, in this context, to note that this is the only instance in which a daylight comet was specifically described by the Chinese as a "broom star" since the year 302. On most occasions, the Chinese either simply state that a comet (also visible at night) was seen in the daytime or that a "star" was seen by day. Significantly, the brilliant daylight comets of 1106 and 1882 were not recorded as having been seen by day in China.

Just 2 days after perihelion, when the comet was only 8 degrees from the Sun, the tail was widely observed as a straight ray rising out of the sunset. March 4 saw as much as 30 degrees of tail, described as being of a bright silver color, streaming away from a planet-like head of second magnitude. Tail lengths of between 40 and 50 degrees were commonly estimated between March 10 and 20 and as far north as Germany, Schmidt could trace it for a full 64 degrees on the 21st, remarking that portions of the tail remained visible hours after the head had set.

The tail was of remarkably uniform (and high) intensity over much of its length, being relatively straight (although a certain degree of curvature was noted far from the head by several observers) and narrow. On March 8, for example, the 43-degree tail widened to just 2.5 degrees farthest from the head, and on the 11th it was judged to be between 0.5 and 0.75 degrees at its widest part, even though its length was in the region of 40 degrees at that time.

Two tails were reported in early March, with the fainter secondary being described as the longer. According to Clerihew in India on March 11, the second (fainter) tail was "twice as long as the first." Unfortunately, he gave no estimate of the actual length of either tail, but other observers on that same evening were making the main tail between 36 and 45 degrees long. If Clerihew was also seeing lengths in the order of 40 degrees, this would imply that the secondary tail was somewhere around 80 degrees long! The two tails formed an angle of some 20 degrees with one another.

The actual length of the main tail has been calculated by various people as being between 1.25 and greater than 2 AU and has even been proposed as the

Fig. 6.4 The Great Comet of 1843 as drawn by C. P. Smyth, Cape of Good Hope, March 4, 1843. Courtesy, B. Warner, University of Cape Town and R. Bouma

longest on record. It was certainly unusually long, but the more extreme estimates may need to be modified if there was a significant curvature along the line of sight. Because we were seeing the tail from a more or less edge-on perspective, this is certainly a possibility.

The comet faded rapidly toward the end of March and was described by E. C. Herrick of Connecticut as being "barely discernible" with the unaided eye on April 3. It was last detected, telescopically, by Smyth and Maclear (Cape of Good Hope) on April 19, when described as being "of the last degree of faintness."

COMET OF MARCH 1843.

Seen from Aldridge Lodge V.D. Land

Fig. 6.5 The Great Comet of 1843 from Aldridge Lodge, Tasmania. Lithograph by Mary Morton Allport (1806–1895). Courtesy, State Library of Tasmania

C/1882 R1

In common with the above entry, the first reports of this comet are only known second hand and originated from outside the astronomical community. Apparently, it was initially seen on September 1 from both the Gulf of Guinea and the Cape of Good Hope. Another early observation was made in New Zealand on the 3rd, and B. A. Gould issued a report from the Cordoba Observatory (Argentina) a few days later, reporting a sighting by an unnamed informant of a bright comet in the morning skies of September 6. The informant had reportedly claimed that the comet was as bright as Venus with a brilliant tail! This was almost certainly a gross exaggeration and probably said more about the conspicuous appearance of the comet than its brightness per se. When experienced astronomers began sighting the comet on the following mornings, the reported magnitudes were far more modest. Thus, W. H. Finlay at the Cape first saw it on the 8th and estimated the central condensation as magnitude 3. An independent discovery was made on that day by Tebbutt in New South Wales, and on the 10th by Joseph Reed on board the *HMS Triumph* off the Cape Verde Islands. Two days later, a further discovery was made by L. Cruls in Rio de Janeiro. Incidentally, the comet

Fig. 6.6 The Great Comet of
1843, March 4, 1843, from
Launceston, Tasmania.
Watercolor by Walter
Synnot (1773–1851).
Courtesy, State Library of
Tasmania

is sometimes referred to in older literature as "Comet Cruls," even though Cruls was just one of many independent discoverers and certainly not the first to see the comet.

By September 15 the comet had become as bright as Jupiter and the tail some 12 degrees long, according to L. A. Eddie at Grahamstown in South Africa. The following day, Gould tracked it all day through a finder scope.

Late on the 16th Universal Time (the morning of the 17th local time), John Tebbutt spied the comet in daylight about 4 degrees west of the Sun and later that same UT day, Eddie watched it rise just 14 min before the Sun and remain visible in daylight with a tail approximately 1 degree long. Also that same day, but on the other side of the world at Ealing in England, A. A. Common was searching for bright comets in the immediate vicinity of the Sun (a course of action that he had been following since learning of the eclipse comet Tewfik the previous May) when he struck the jackpot!

Very interesting observations of the comet were made late in the day by Finlay and Elkin at the Cape of Good Hope. Using an early type of solar filter on a 15-cm (6-in.) telescope at 110 magnification, Finlay watched the comet rapidly approach the limb of the Sun. The comet was described as being silvery in color (contrasting with the reddish-yellow Sun) and was measured with a micrometer as just 4 s of arc across, with a very short tail. Observing through another telescope, Elkin "actually observed it to disappear among the undulations of the Sun's limb" at September 17.6506 UT, and Finlay himself saw it vanish 8 s later "when the sun's limb was boiling all about it." He possibly caught a quick glimpse of it again 3 s afterwards, but he could not be sure. After the comet had disappeared, Finlay searched the Sun's face for a possible trace of a transit, but found no evidence of the comet.

Just how bright might the comet have been at the time of these observations? From Elkin's and Finlay's descriptions, it would seem that the surface intensity of the tiny but brilliant head was little inferior to that of the Sun's limb itself. Finlay also measured its diameter as 4 s of arc. Now, assuming the comet's head and the Sun's surface were of equal intensity, the difference in brightness can easily be computed from their difference in surface area. On this assumption, the comet would be about 202,000 times – almost 13.3 magnitudes – fainter than the Sun – in other words, magnitude –13.4. However, Finlay and Elkin saw the comet adjacent to the *limb* of the Sun, not the full-on face of the solar photosphere. This is significant, as the well-known phenomenon of limb darkening reduces the intensity of the solar limb by about 60%, i.e., it is only 40% as bright as the center of the solar disk. Taking this into consideration, the comet's brightness rounds out at close to 12.5. Much brighter, and it might have very briefly stood out as a bright spot against the darker solar limb!

As a matter of interest, Elkin compared the comet's transit of the Sun with the occultation of a fourth magnitude star by the full Moon. A direct comparison of magnitudes would then make the comet somewhat fainter than our estimate (around –10), but this was probably an impression of ease of visibility rather than a literal magnitude estimate. We put forward 12.5 as a reasonable estimate for the comet at the time of the transit, using the Sun as the comparison star!

The transit ended on September 17.69, and the comet arrived at perihelion at 17.72, just 0.00775 AU from the Sun. No observations were made at the time of perihelion, and the comet, still unseen, reached a maximum solar elongation of 27 min of arc on 17.74 before moving back toward the Sun again and passing directly behind it at 17.79. It remained occulted until 17.87 and does not seem to have been observed again until September 18.06, when Tebbutt found it in broad daylight less than 1 degree from the Sun's western limb.

Later observations strongly suggest that the intrinsic brightness of the comet surged at or soon after perihelion, probably due to the disruption of the nucleus about which we shall speak below. If the surge began at perihelion passage itself, the comet may have been close to –12 when Tebbutt saw it. In any case, it was unlikely to have been fainter than –9, fully justifying Tebbutt's comment that the comet was by far the brightest that he had ever seen.

Later that (UT) day, David Gill saw the comet rise pure yellow from behind the mountains on the eastern side of False Bay, South Africa, and remain visible throughout the daylight hours as a brilliant object with a tail about 0.5 degrees long.

Incidentally, in connection with Gill's observation, the following passage was found in a book on comets published several decades ago and repeated in more recent works. Purportedly written by Gill, it reads:

There was not a cloud in the sky, but looking due east one saw the tail of the Comet stretching upwards, nearly to the zenith, and spreading with a slight curve. Not a breath stirred; the sky was a dark blue almost to the horizon. The scene was impressive in its grandeur. As the Comet rose, the widened extremity of its tail extended past the zenith and seemed to overhang the world. When dawn came, the dark blue of the sky near the point of sunrise began to change into a rich yellow, then gradually came a stronger light, and over the mountain and among the yellow, an ill-defined mass of golden glory rose, in surroundings of indescribable beauty. This was the nucleus of the Comet. A few minutes later the Sun appeared, but the Comet seemed in no way dimmed in brightness, and although in full sunlight the greater part of the tail disappeared, the Comet itself remained throughout the days easily visible to the naked eye; with a tail about as long as the moon is broad.

Beautiful and well written, but there is one big problem. Except for the final sentence, it is all baloney! For one thing, the length of tail implied here is a physical impossibility.

What Gill actually wrote was:

I was astonished at the brilliancy of the comet as it rose behind the mountains on the eastern side of False Bay. The Sun rose a few minutes afterwards, but to my intense surprise the comet seemed in no way dimmed in brightness, but becoming instead whiter and sharper in form as it rose above the mists of the horizon. I left Simons Bay and hurried back to the observatory, pointing out the comet in broad daylight to the friends I met by the way. It was only necessary to shade the eye from direct sunlight with the hand at arm's-length to see the comet with its brilliant white nucleus, and dense white, sharply bordered tail of quite ½ degree in length.

John Bortle points out that Gill actually did observe a very long tail for the comet of 1880 (see above), albeit pointing eastward in the evening sky, and he suggests that somebody may have run together his accounts of both comets to produce the above spurious description.

Exaggerated accounts aside, the comet nevertheless became an incredible sight during late September. There were numerous naked-eye daylight sightings on the 18th and 19th, and even as late as September 22, E. E. Barnard (Nashville) followed the comet with the unaided eye for 15 min after sunrise. Prior to sunrise, he noted 12 degrees of tail in the morning twilight. Only one previously recorded comet – that of 1402 – remained visible with the naked eye in daylight for a longer period of time!

Also on the 22nd it was observed telescopically just before noon by J. M. Schaeberle in Michigan and just after noon by E. Millosevich in Rome, and during the following days it remained a brilliant spectacle in the morning twilight. On September 27, after stars of the first magnitude had vanished in the rising dawn, the comet and 10 degrees of tail remained visible with the naked eye.

The end of September and early October saw the comet still close to zero magnitude and sporting a brilliant 20-degree tail, with a dark lane running down the middle like a shadow of the nucleus. The tail was remarkably narrow, being estimated as just 1 degree wide in late September.

At the beginning of October, a number of observers reported the central condensation as elongated and definitely split on the 3rd, according to F. Terby

(Leuven, Belgium). Three days later, observers were reporting three distinct nuclei, and on October 15 Eddie (Cape of Good Hope), using a 24-cm (9.5-in.) reflecting telescope, noted one distinct nucleus that "resembled in color the electric light" in addition to two more within a "bar of light." Moreover, when the magnification was increased to 100, the condensations within this bar "seemed again doubled, so that the whole nucleus resembled a string of five ill-defined luminous beads." This multiple nucleus remained visible throughout the remainder of the comet's apparition.

Also during October, the comet's head region became enveloped in a fainter veil of light extending sunward and at times appearing as a "tube of light," with the inner portion darker than the edges. The sunward projection was at least 4–6 degrees and bright enough to be apparent with the naked eye.

An even more peculiar phenomenon was noted during the same month. As early as October 5, Markwick observed two "wisps" preceding the comet's head, which he seemed inclined to dismiss as background nebulosities, except that he could find no nebulae at that location on his star charts. Then, 3 days later, J. Schmidt discovered a "comet" about 4 degrees southwest of the Great Comet and duly reported his discovery. He also saw the same object (or what he presumed to have been the same object) on October 10 and 11, and from these positions rough orbits were calculated that suggested a "sunskirting" perihelion around September 25.

E. Hartwig (Strasbourg, France) also saw this, or a similar, object on the 10th and described it as "a comet with a bright nucleus and a fan-shaped tail." He tried for the object again on the 13th but without success.

Nevertheless, the following day Barnard found a nebulous mass some 15 min of arc in diameter to the south of R1. Not far from this nebulosity, he saw a similar object apparently in physical contact with the first and, on its opposite side, a fainter third one. Sweeping his telescope, he located several more objects (one of which appeared very elongated), counting 6 or 8 of them within 6 degrees of the large comet's head. Later, on October 21, W. R. Brooks found a comet-like object some 2 degrees long, situated 8 degrees east of the main comet.

What were these objects?

They certainly looked like genuine comets, and it is tempting to think of them as fragments that separated from the main nucleus either at perihelion or much earlier, while the comet was still far from the Sun. Each of these suggestions runs into trouble, however. To be several degrees away from the main comet only weeks after perihelion, the velocity of separation (had this occurred near perihelion passage) would need to be very high and indicative of something far more violent than the splitting of the nucleus itself. Also, why were there no observations earlier when the "comets" would presumably have been brighter, more condensed, and closer to the parent object?

But if the fragments were already separate from the parent prior to perihelion, how could such small bodies survive their encounter with the Sun? Or, if they were large enough to survive perihelion initially, they would surely have appeared as bright secondary comets surrounding the primary one. So why were they not seen?

Gary Kronk's suggestion that these objects were not so much "comets" in the true sense of evaporating icy bodies as condensations within the sheath of light surrounding C/1882 R1 seems a plausible one. Perhaps (and this is just speculation) they resulted from the disruption of concentrations of large particles within the sheath in a similar manner to the striae produced by particle disruption in normal dust tails.

The comet remained conspicuous throughout October and November. According to B. J. Hopkins (London), the "nucleus" was of second magnitude on November 9, although it is probable that he was using "nucleus" very loosely as essentially synonymous with "head." On the 12th the brightness of the tail close to the head was said to have corresponded to out-of-focus images of third-magnitude stars. Probably, the head itself would not have been much brighter. Markwick commented that, on November 21st, the comet was "superior in brightness" to a fifth-magnitude star. How "superior" he did not say, but the fact that he used a star of that magnitude for his comparison hints that the comet was probably "inferior" to one of fourth magnitude. The comet was the subject of a series of spectacular photographs by David Gill during November, and it is possible to compare it with some stars of known magnitude on these images. These comparisons indicate a brightness of about 3.5 in the middle of the month.

Fig. 6.7 The Great Comet of 1882, photographed on October 13 from Melbourne Australia, by T. W. McAlpine. Courtesy, State Library of Victoria

Fig. 6.8 The Great Comet of 1882, November 7, 1882, photographed by Sir David Gill. Courtesy, of the SAO

The comet was magnitude 5 on December 1, according to Markwick, yet 15 degrees of tail could still be seen without optical aid and despite moonlight. Remarkably, naked-eye reports continued until as late as March 8, 1883, when Gould glimpsed it without optical aid for the final time. In view of the intensity of the tail, it is possible that the late naked-eye observations may have been of the tail rather than of the head of the comet. The final telescopic view of the comet was on June 1, when Thome described it as "an excessively faint whiteness."

Perhaps the final word on this comet should be left to Anna Richards, who decades later wrote in a letter to *Sky & Telescope*:

> I have for 75 years been intensely interested in the night skies. I have observed from mountains, the desert, the deep woods and even from the ocean. I have witnessed many unusual phenomena, but the glorious comet of the early '80s was by far the greatest of all.
>
> My impressions are those of a little child, but very clear and vivid. The first appearance was in the early Fall, and it was visible all the following winter. There were no radios and very few newspapers, but when I heard people talking about something new in the sky I was all interest. It was some time before I could locate the comet because the brilliant Colorado sunshine kept it dim at first. But finally its splendor filled the sky.

Fig. 6.9 The Great Comet of 1882 on the morning of November 14, 1882. Photograph by Sir David Gill. Courtesy, of the SAO

Fig. 6.10 Photograph of the Great Comet of 1882 by an unknown photographer. Courtesy, State Library of Victoria

Fig. 6.11 The Great Comet of 1882 from Greenwich, MA. Painting by B. Brooks. © Stuart Schneider from *Halley's Comet, Memories of 1910*, by S. Schneider, and *Wordcraft.net*

We were high in the Rockies west of Denver. Our view was entirely unobstructed in that clear atmosphere. My work took me back and forth each night and morning while it was dark. When the deep snows of winter covered the earth, with cliffs and evergreen trees to break the expanse of white, it was then the comet shone brightest. Its length seemed to reach over one fourth of the sky.

The comet was visible so long that we began to regard it as a permanent fixture. As the day grew longer I forgot the comet for a time. When I remembered it I scanned the sky, but in vain. It was gone.

C/1965 S1 (Ikeya–Seki)

Unquestionably the most brilliant comet of the last century, this majestic object is the subject of many fond memories by older generations of comet observers and has even been commemorated in a musical composition. For southern observers at least, it has acquired an almost legendary aura and has become a sort of unofficial "standard" against which other great comets are measured. A browse through astronomical forums on the Internet will find, in more than one location, the question "Was it [Hyakutake, McNaught, or whatever] as spectacular as Ikeya–Seki?" to which the answer is generally given as "No!"

Nothing hinting of its future glory was apparent, however, when the comet was discovered as an eighth magnitude tailless ball of nebulosity by K. Ikeya and, independently, T. Seki on the morning of September 18, 1965. Both amateur astronomers had two comet discoveries already and, remarkably, were destined to share another comet in late 1967.

A very preliminary orbit gave no cause for excitement, suggesting a perihelion distance near Earth's distance from the Sun, but as further measurements of the comet's track were made, it became apparent that this initial attempt was wide of the mark. By late September, it had become clear that the new comet was another Kreutz sungrazer with an orbit very similar to that of 1882. An early sungrazing orbit actually had the comet hitting the Sun, but further refinements showed that it would reach perihelion at 0.008 AU on *October 21.18*. The striking similarity between its orbit and that of C/1882 R1 was confirmed.

The comet brightened perceptibly from day to day, and by September 26, a tail of 1.5 degrees length was photographed by James Young at Table Mountain, California. The naked-eye limit was reached at the very end of September, and the tail continued to develop, with typical plasma rays photographed by Elizabeth Roemer at Flagstaff on October 7. Observing from near the CBD of Sydney, Australia, New South Wales government astronomer Dr. Harley Wood sighted the comet, first through a small telescope and then with the naked eye, on October 11. Continuing to brighten, it was estimated as magnitude 0.2 on October 16 according to Arthur Page at Brisbane, Australia, and 0 on the 18th as estimated by astronomers at Woomera in South Australia. Up to 10 degrees of tail were visible by mid-month.

Writing in the October 1965 issue of *Sky & Telescope*, Dr. L. E. Cunningham remarked that the comet's brightness in early October was "about the same as [C/1882 R1] under similar conditions" and added that "Observability will be the best possible for this sungrazing group." We may think that both these statements were overly optimistic, but Cunningham was probably thinking of the early estimates of C/1882 R1 by Finlay on September 8th that year. As we saw, C/1882 R1 apparently surged in brightness and activity after the disruption of its nucleus at perihelion, and it was hoped that a similar fate might befall Ikeya–Seki. Up to a point, this did indeed happen, as we shall see, but the surge in activity was neither as great nor as long lasting as that of the earlier comet. Despite the magnificent display that it provided, C/1965 S1 was clearly a less massive comet than its famous predecessor.

As for Cunningham's statement that the observability of the 1965 comet would be the best possible, he was obviously referring to the period very close to perihelion. Unlike the comets of 1843 and 1882, Ikeya–Seki was neither eclipsed by the Sun nor passed in transit across its face, but could be viewed throughout its entire encounter. This was, indeed, the best possible view, but apart from this very brief period near perihelion, the comet remained on the far side of the Sun as seen from Earth and suffered a degree of foreshortening of the tail. In that sense, it was not as favorably placed as the comets of 1880 and 1887, which were closer than the Sun and whose tails pointed more or less in our direction.

Ikeya–Seki became a naked-eye object in broad daylight on October 20. As observed by G. de Vaucouleurs at McDonald Observatory in Texas, it had a magnitude of –10 and a 1–2 degree tail at local noon (18:00 h UT). Half an hour later, Norbert Roth and Darrell Fernald at the Smithsonian Station at Organ

Pass, New Mexico, estimated the tail to be 1 degree long and as intense as the 25 and half a day old Moon, also visible in the daytime sky. Two hours later, Elizabeth Roemer estimated the brightness of the comet as –10 or –11 and the tail length as 2 degrees, as seen from Flagstaff, AZ.

As the comet drew closer to perihelion, numerous naked-eye daylight sightings were made from around the world. Several people even photographed it with ordinary box cameras simply by screening the Sun behind a foreground object. One such photograph showed the comet and about 1.5 degrees of tail shining through the branches of a tree. The Sun was simply obscured by one of the branches!

At the coronagraph station of Tokyo Observatory, located on Mount Norikura, the comet was described as "10 times brighter than the full moon" just half an hour before perihelion. This description implies a magnitude of around –15! A disruption occurred immediately prior to perihelion passage, and the comet was seen to divide into three separate pieces shortly after passing through perihelion. Later, only one piece remained visible.

The comet was again visible with the unaided eye in daylight on October 21 and telescopically the following day. Pre-sunrise observing began on the morning of the 23rd, when it was located in bright twilight at Pretoria, South Africa. On the morning of October 25, observers at the Smithsonian Station in Arequipa, Peru, saw a tail 20 degrees long and 3 degrees wide at the remote end, tapering from an almost star-like head of magnitude –2.

Tail length estimates of up to 30 degrees (corresponding to actual dimensions of 0.75 AU) were commonplace during the last days of October and early November. The greatest estimate was probably that of R. B. Minton, who traced it for possibly 45 degrees on the morning of October 28. Incidentally, the sometimes-quoted length of 60 degrees by Richard Nelson, observing from the Tehachapi Mountains of California on the 31st resulted from a misreading of Nelson's report (quoted in the January 1966 issue of *Sky & Telescope*). The 60-degree estimate actually related to the cone of the zodiacal light on that morning, not the comet's tail. The latter was estimated by Nelson as 20 degrees naked eye and 23 degrees as observed through a pair of binoculars.

Regarding Minton's October 28 observation, it is interesting to note that he also saw and photographed a small sunward-pointing anti-tail on that morning, in addition to a faint and straight secondary tail emerging from the southern (convex) edge of the main one. This latter feature was probably a plasma tail and was also present in photographs taken on October 31 and November 1 by Bradford and Tamie Smith at New Mexico. The Smiths' photographs show at least two very faint rays within the secondary tail that, if extended, appear to converge at or very near the comet's head, in contrast with the brighter striae within the main (dust) tail. The rays and the closest striae form angles of about 25 degrees with each other, the striae extending more or less obliquely across the tail. It is the presence of these striae that gave the main dust tail its characteristic cork-screw appearance of late October/early November.

An interesting feature photographed by M. Bester at Boyden Observatory (South Africa) on November 4 was also probably related to the secondary plasma tail. About 4 degrees from the head, a very faint "finger" of nebulosity extends approximately 1.7 degrees at an angle to the main tail (measuring from the direction of the head) of around 160 degrees. This feature looks somewhat ragged and gives the appearance of being partially detached from the tail itself. My guess is that this is a contorted, possibly disconnected, extremity of the plasma tail and may well be evidence that a disconnection event took place early in November. The brilliance of the main tail, however, obscured further evidence of this.

The main tail was certainly intense and, although the comparison could not have been made in 1965, reminds one (on looking back) of a laser beam. Personally, my best view was on the morning of October 31, when the tail tapered out from a small but intense head of about magnitude 2.5 and maintained an almost constant intensity over a length of some 20 degrees. Writing in his observing log on that same morning, John Bortle noted that a light mist steadily thickened into a fog until the sky became so opaque that stars fainter than magnitude 2.5–3 were invisible to the naked eye. Yet, the tail of the comet was neither diminished in intensity nor reduced in length, but continued shining through the fog without hindrance.

Fig. 6.12 Comet Ikeya–Seki from Canberra, October 31, 1965. Painting by David Nicholls ©1965, 1997 David Nicholls, used with permission

Fig. 6.13 Comet Ikeya–Seki, November 7, 1965. Courtesy, David Nicholls

Photographs of the tail around that time revealed a brighter inner core and, at the very center, a very narrow dark lane like the shadow of the nucleus, very similar to the feature described in the comet of 1882. As the comet faded during early November, the intensity of the tail scarcely changed for a period of more than a week. Observing on November 2, Bortle attempted to measure the intensity of the tail in terms of extrafocal star images at various distances from the head. At just a few degrees, he found the tail matched out-of-focus images of second magnitude stars, at 8 degrees (where the tail was still less than one degree wide), it matched images of third-magnitude stars and near the terminus (about 20 degrees from the head), stars of magnitude 4.5. He also noted that, unlike most comets, the tail ended suddenly rather than slowly fading off into the night sky. At that time, the very small and tightly condensed head was around magnitude 3, little more than a point of light at the sunward extremity of the brilliant tail.

While these observations were being made of the tail, other astronomers, concentrating on the comet's central condensation, found that it appeared distinctly elongated in early November and clearly double on the 4th, according to Howard Pohn at Flagstaff. A third condensation was also observed, briefly, further along the tail and other possible ones reported by several astronomers around the same date. Only the main pair remained visible, however, and these persisted until the final observations in January. Interestingly, the secondary condensation itself showed as a close triple on the original negative of a photograph by A. D. Andrews at Boyden taken on November 6. The breakup is believed to have occurred close to perihelion, probably about the time Japanese astronomers noted a disruption within the comet's head, although the fragments that they watched separate at that time were apparently not the ones observed in November.

From the southern hemisphere, the comet remained a naked-eye object throughout November. This author saw it on the morning of November 25 and for the first time could see the head as an object of some dimension, rather than merely a bright point at the sunward extremity of the tail. It resembled the globular star cluster 47 Tucanae. On that same morning, John Davies, observing from the Blue Mountains west of Sydney, judged the brightness to be 3–3.5 with as much as 30 degrees of tail visible naked eye. Photographs at that time recorded tail lengths as great as 35 degrees. Even as late as December 5, the tail could still be detected with the naked eye as a faint, but clearly visible, streak of light. The head was not visible without binoculars, appearing as a diffuse and transparent globule of magnitude 7 or thereabouts. Just 10 days later, all that remained visible was a very diffuse patch of light, possibly 20 min of arc in diameter as observed with 20 x 65 binoculars, but very vague and ill-defined. Not the slightest trace of tail could be detected. The last positive observation of the comet was on January 14, 1966, although a faint image may have been recorded by Baker-Nunn cameras as late as February 12. Based on the comet's brightness before perihelion, it should still have been visible in small telescopes in early February, and, from southern observations in late November, may even have been accessible in good binoculars. Clearly, the comet faded very rapidly after November.

Thus far, Ikeya–Seki remains the last of the great Kreutz sungrazers to grace our skies. Yet, if Sekanina and Chodas are correct in their latest analysis of this fascinating group of comets, there may be other Ikeya–Seki's in the future, possibly not too far ahead. This raises the question: Will the *next* comet rated as among history's greatest be another sungrazer? Let's hope that we will not need to wait long for the answer . . . whatever that answer might be!!

Chapter 7
Daylight Comets

Several times each century, a comet shines brightly enough to be visible even when the Sun is above the horizon. From the records of the past couple of centuries, it would appear that this beautiful phenomenon is not as rare as a reading of historical records might suggest, and that many daytime comets of earlier years have gone unnoticed and unrecorded.

It may seem ironic that many of the most brilliant comets of history were probably missed when at their peak brightness, but there are actually very good reasons why this might have happened. Because the intrinsic brightness of a comet is so strongly dependent on its distance from the Sun (as explained in Chapter 1), most comets capable of reaching daylight brilliance come to perihelion well within Earth's orbit and, for the brief time that they shine at their brightest, appear at very small elongations, i.e., very close to the Sun in our skies. The very brightest ever recorded were seen at elongations of less than 3 or 4 degrees. Such comets as these rivaled the magnitude of the full Moon, yet even this pales in comparison to the brilliance of the Sun!

The phenomenon of forward scattering of sunlight from small particles of dust can greatly enhance the apparent brightness of a comet located between Earth and the Sun, as shown most dramatically by C/McNaught in mid-January 2007. Unfortunately, though, forward scattering not only applies to the dust of comets. It works equally for the dust motes in our own atmosphere. A dusty atmosphere (and the atmosphere is always dusty to some degree!) will therefore give a bright aureole around the Sun in just the position where bright daylight comets are typically located. Any degree of haze at all will therefore make it very difficult to find a daylight comet.

Yet, in spite of these difficulties, some daylight comets did find their way into ancient records. But here we strike another difficulty! Sometimes the record itself is difficult to decipher; does it *really* imply daylight visibility or not?

For example, Chinese records of the comet of May 147 B.C. note that it "moved away at dawn and became smaller." What exactly does this mean, though? Does "dawn" mean "sunrise" (as sometimes it can), or does it simply mean twilight? We assume here that it just means twilight and that this was not a daylight comet, although we can't be certain about this.

D. Seargent, *The Greatest Comets in History*,
DOI 10.1007/978-0-387-09513-4_7, © Springer Science+Business Media, LLC 2009

Sometimes statements in secondary and tertiary sources can be misleading and accounts are recorded that have more in common with the notorious urban legend than with accurate reports! This may be true of the inclusion of the great comets of 1532 and 1618 in George F. Chambers' list of daylight comets in his 1909 book *The Story of the Comets*. Neither comet was likely to have been sufficiently bright for daylight observation.

A reading of Vsekhsvyatskij's account of the comet of 1737 seems to imply that this was visible in the daytime. According to this author, the comet was found "in sunlight" from Lisbon on February 9. Taken at face value, this would appear to suggest daylight visibility, but the original wording of the account quite clearly places the comet in evening twilight, not in full daylight.

Other accounts, though apparently primary, appear equally misleading. For instance, a description of the comet of 1315 by French physician and astrologer Geoffrey of Meaux states that this comet was "visible day and night" without setting from December 1315 through February 12 the following year. Intrinsically one of the brightest known (absolute magnitude -3 according to Kronk!) but with a perihelion distance of around 1.65 AU in mid-October 1315, this comet could hardly have been bright enough for daylight visibility at the end of the year. Or at any time for that matter! Geoffrey's statement appears to have simply been a hyperbolic way of saying that the comet was circumpolar.

In a similar vein, mention is made in R. Knox-Johnson's *The Cape of Good Hope, A Maritime Journey* (1989) of a sighting by Pedro Cabral of "a comet … so bright … that it was visible day and night" for a period of 10 days in May 1500. Mention is also made of this comet in *The Voyage of Pedro Alvares Cabral to Brazil and India* (1938), where it is said to have possessed "a very long tail in the direction of Arabia." This appears to have been the same comet as seen in China from early May until July of that year and which has been proposed as a good candidate for the previous return of Tebbutt's Comet of 1861. Tebbutt's was visible in daylight under exceptional circumstances in 1861, but neither the Tebbutt orbit nor the Hind nor Hasegawa orbits for the 1500 comet suggest daylight visibility, and it seems that a literal reading of Cabral's statement is doubtful.

Published accounts of C/1857 Q1 (Klinkerfues) refer to two (telescopic?) observations by Reslhuber on October 4 and 5 as having been made "just before sunset." Nevertheless, the exact times of day recorded were a little *later* than his October 3 observation, which appears to have been in bright twilight! The comet is unlikely to have been brighter than third magnitude at the time, although it would have been strongly condensed.

On the other hand, the "sun stars" listed by Strom from Chinese records have been included here, although their nature is open to question, as we saw in Chapter 6.

Not surprisingly, several of the objects listed below have already been dealt with at some length in Chapters 1–6 of this book. As it would be redundant to repeat a detailed description of these, they are simply mentioned here. Brief details are, however, given for those objects that have not been covered earlier.

THE COMET, AS SEEN AT 6.15 P.M., ON JUNE 18.

Fig. 7.1 Comet Wells, June 15, 1882, from Melbourne, Australia; wood engraving by Alfred Martin Ebsworth. This comet was telescopically visible in daylight near perihelion from June 10–12. Courtesy, State Library of Victoria

Fig. 7.2 Comet Kohoutek, January 11, 1974, photographed from the Catalina Observatory by a team from the University of Arizona. Although never especially bright in the night sky, this comet was observed in daylight through binoculars when near perihelion on December 28, 1973. Courtesy, NASA

Fig. 7.3 Comet McNaught visible against a blue daytime sky January 12, 2007. Courtesy, Terry Lovejoy

185 B.C.

The Chinese text *Han Shu* notes that "In the autumn, a star appeared in daytime." This may have been Venus, which was 20 degrees from the Sun at the time.

182 B.C.

The same Chinese text notes that "in spring, a star was visible in daytime." Again, Venus is a possibility as it would then have been 40 degrees from the Sun and near magnitude -4.

104 B.C.

In the *Book of Prodigies after the 505th Year of Rome* (written around the 4th Century A.D.), Julius Obsequens writes that "The moon and a star appeared by day from the third to the seventh hour" sometime in 104 B.C. This may have been Venus, but the record's reliability may also be open to question.

44 B.C. (C/-43 K1)

"Caesar's Comet"; see Chapter 3 for an account of this object.

15

The Chinese text *Han shu* notes that a "star" was visible at noon sometime during this year. Strom also notes a sun star observation between March 10 and April 7.

302

Oriental records mention a comet seen in the daytime in May or June (see also Chapter 6).

363

Roman historian Ammianus Marcellinus wrote that "in broad daylight comets were seen" late that year. The Chinese record a comet in August, in the evening sky, but give no indication of great brightness. I. Hasegawa cites the Chinese comet of 363 as a likely candidate for the previous return of Comet Bennett, the Great Comet of 1970 and suggests perihelion in late August. If it was Bennett, the 363 return would have been a good deal less favorable than that of 1970 and the comet less bright. Yet, in 1970 Bennett was far from being bright enough for naked-eye daylight visibility. It was not even observed telescopically in the daytime, although M. J. Hendrie managed to hold it in telescopic view until just 2 min 30 s before sunrise in early April.

Another comet, visible only in daylight, may have appeared in late 363. A Kreutz sungrazer appearing very late in the year would have had a strong southerly declination and might have been seen from Italy only in the daytime close to perihelion. Although just wild speculation, could the reference to "comets" – in the plural – imply several sungrazing fragments close together? The account is, however, just too brief to know.

1077

An English record mentions a "blazing star" near the Sun on April 9. Pingre suggested that this may have been Venus, but it has also been suggested that it might have been a Kreutz sungrazer close to perihelion.

1066 1P/Halley

Halley's Comet may have been observed in daylight during the 1066 return (see Chapter 2).

C/1106 C1

Bright daylight comet observed near the Sun in early February (see Chapter 3 for details of this object).

1222 1P/Halley

Korean records note that Halley's Comet was visible during the daytime on September 9 (see Chapter 2 for discussion).

C/1402 D1

The Great Comet of 1402, visible in daylight for 8 days in March. A single record of a supposed second daylight comet in August is believed to be a misdated account of C/1402 D1 (see Chapter 4 for details).

C/1471 Y1

The Great Comet of 1472, said to have been visible "even ... at midday" on January 23 (see Chapter 4 for details of this comet).

1539

According to R. Strom, the Chinese recorded a sun star some time between July 15 and August 13. A comet was also recorded in April and May of 1539, but could not have been associated with the sun star.

1564

Another sun star recorded some time between August 7 and September 4.

C/1577 V1

The brightest comet of the sixteenth century, discovered by Tycho Brahe before sunset on November 13 as a star-like object as bright as Venus. The long tail became visible after the Sun set (see Chapter 4 for an account of this comet).

1587

A star was seen throughout the day on August 30, according to Japanese records. Unlikely to have been associated with the alleged Korean comet of October that year (see Chapter 4 account of C/1577 V1 for more details of the Korean record).

1625

A sun star was recorded on September 2. This object was said to have been "at the Sun's side."

1630

Another sun star recorded on August 5.

1643

A sun star was seen sometime between June 16 and July 15. Strom notes that two sun star records exist for 1643.

1644

Another sun star was recorded sometime between March 9 and April 6.

1647

Yet another sun star recorded on July 28.

1648

According to Sir Richard Baker's *A Chronicle of the Kings of England* (1684), "a Star appeared visibly at Noon, the Sun shining clear" on May 29, at the time "the King rode to St. Paul's Church to give thanks for the Queen's safe delivery of her second Son Prince Charles." This may have been a daylight meteor, although the Chinese also recorded a sun star in the summer of that year.

1665

A sun star was recorded on August 27. The Great Comet of 1665 (C/1665 F1) was not seen later than April 20 (perihelion at 0.1 AU on April 24) and could not have been associated with the sun star.

C/1680 V1

The Great Comet of 1680, visible in daylight near perihelion (see Chapter 4 for an account of this comet).

C/1743 X1

The Great Comet of 1744, sometimes known as Comet de Cheseaux or Comet Klinkengerg–de Cheseaux, visible both telescopically and with unaided eye in broad daylight for several days near perihelion (see Chapter 4).

1774

A sun star was seen in China sometime during this year.

1792

A sun star was seen sometime between April 21 and June 20.

1839

A sun star was seen on August 14.

C/1843 D1

The "Great March Comet" was almost certainly the brightest since 1106 and was probably the most conspicuous (though not necessarily the brightest) daylight comet on record (see Chapter 6 for an account of this great Kreutz sungrazer).

C/1847 C1 (Hind)

A telescopic object when discovered by J. R. Hind (London) on February 6, this comet reached naked-eye visibility in early March and displayed 4–5 degrees of tail before entering deep twilight after the middle of that month. Perihelion (0.04 AU) came on March 30.78 and Hind and colleagues were able to observe the comet telescopically in daylight near that time. Hind described it as having a "nucleus" that "was round or nearly so" with "Two short rays of light (which) formed a divided tail, not more than 40" in length." Hind noted that the comet was also seen in daylight on the same day by observers in Truro and Ynys Mon and "about noon, by a clergyman residing in the Isle of Anglesey." It seems that all of these observations were telescopic, making this the first daylight comet observed purely with optical aid.

C/1853 L1 (Klinkerfues)

This comet was discovered by W. Klinkerfues (Gottingen) on June 11 as a relatively faint telescopic object, but brightened to naked-eye visibility around August 3 and had become a bright object showing some 12.5 degrees of tail by August 28. J. F. J. Schmidt located the comet in daylight with a 13-cm (5.2-in.) refractor on August 30 and followed it until September 4. It was also seen, telescopically, by J. Hartnup at Liverpool on September 3. Perihelion (at 0.31 AU) occurred on September 2.2. No naked-eye daytime observations were made.

C/1858 L1 (Donati)

This great comet was observed by several astronomers before sunset, but only with the aid of telescopes (see Chapter 5 for details of this comet).

C/1861 J1 (Tebbutt)

On June 30, as the comet passed close to Earth, it was plainly seen with the naked eye prior to sunset.(see Chapter 5 for discussion of this comet).

1865

A sun star was recorded in China on July 18. The Great Comet C/1865 B1 arrived at perihelion (0.03 AU) on January 14 and could not have been associated with the sun star.

C/1882 F1 (Wells)

This comet was discovered by C. S. Wells (Dudley Observatory, New York) as a telescopic object, on March 18. Perihelion (0.06 AU) occurred on June 11.03 and the comet reached naked-eye visibility in late May, rising to about magnitude 0, with 5 degrees of tail visible, before entering morning twilight at the end of the first week of June. Daytime observations were made by several people using a variety of instruments between June 6 and 12, after which the comet moved into the evening twilight and was observed from the southern hemisphere as it quickly faded. Spectroscopic observations near perihelion revealed the D lines of sodium, the first time that this element had been observed in a comet.

C/1882 S1

The "Great September Comet" was one of the brightest ever recorded and remained visible with the naked-eye in daylight for a longer period than any other known comet except that of 1402.(see Chapter 6 for discussion of this Kreutz sungrazer).

1882

On December 21, 1882, several people at Broughty Ferry (Scotland) saw a bright object near the Sun between 10 and 11 a.m. local time. The star was said to have had "a milky white appearance" unlike the "brilliant luminous radiance" of stars at night. Through a "glass", it was said to have shown a crescent shape. On December 25, a correspondent to the *Dundee Advertiser* (which carried the initial report) identified the object as Venus, but in the 1883 January 5 issue of *Knowledge*, J. E. Gore argued that Venus would have been too far from the Sun at the time to account for the report.

X/1896 S1

As L. Swift (Lowe Observatory, CA) watched the Sun set behind a spur of the Sierra Madre range on September 21, 1896, his attention was drawn to a bright object about 1 degree above the partially set Sun. The object appeared about as bright as Venus and, through an opera glass, was found to be accompanied by a second and much fainter one some 30 min of arc to its north. The following evening, Swift attempted to re-locate the objects with an 11-cm (4.3-in.) telescope, but only managed a fleeting, 5-s glimpse of one of them at a time when the Sun was half below the horizon. This object was faint in comparison to the previous afternoon and Swift suspected that it was the smaller companion. It is, I think, more likely that this was the main object greatly faded. Presumably the companion was then too faint to be seen.

The most likely explanation for these observations is a comet, approaching and receding from behind the Sun, which had split some time prior to perihelion. It is possible (though, I think, unlikely) that the object glimpsed by Swift on the second night was a *third* component.

C/1901 G1

The Great Comet of 1901, sometimes known as Comet Viscara, was discovered as a naked-eye object with a short tail on April 12, 1901, when located in the morning sky. With perihelion (0.24 AU from the Sun) on April 24.75, the comet became a very striking morning object from the Southern Hemisphere before dawn prior to perihelion, and an initially spectacular evening object as it faded through May.

On April 25, R.T.A. Innes and J. Lunt at Cape Town followed the comet through a 46-cm refractor "for some time after sunrise" when the nucleus was estimated as being as bright as Mercury. It was also visible in daylight through the 25-cm (10-in.) guiding telescope.

That same day, W.E. Cooke and C. Todd at Adelaide in South Australia followed the comet for some time after sunrise using a 20-cm (8-in.) telescope.

On the other hand, the report that it was seen at Yerkes Observatory for 15 min after sunrise of April 24 seems to be spurious.

C/1910 A1

The "Great January Comet" of 1910 was observed both telescopically and with the naked eye in broad daylight near the time of perihelion (see Chapter 5 for a description of this comet).

1P/Halley

This famous comet was observed telescopically in daylight near perihelion and both telescopically and with the naked eye near the time of closest approach to Earth nearly 1 month later (see Chapter 2 for an account of the spectacular and unusually bright 1910 return).

1921

A star-like object about the same brightness as Venus was observed near the setting Sun on August 7 by a group of Lick Observatory astronomers including H. N. Russell and W. W. Campbell. The object shared the motion of the Sun and was clearly astronomical.

Other observations of bright objects in the twilight or daylight sky were received around the same time. Two observers in England saw what appears to have been the same object at sunset about 8 h prior to the Lick observations. Comparing the English and Lick positional estimates, it seems that it moved a degree or two closer to the Sun during the intervening period.

Other reported observations were not consistent and were probably sightings of one or another of the bright planets. In particular, a sighting from Germany only half an hour after the English ones implied a rate of motion of 27 degrees in just 8 h! This not only strongly contradicts the far slower movement implied by comparing the English and Lick observations but (according to astronomer Max Wolf) would have placed the object just 0.005 AU from Earth. Though not impossible, this is highly improbable. On the other hand, the position of the "German" object turned out to be very close to that of Jupiter and, as no mention was made of that conspicuous planet, it seems safe to conclude that the "German" object was none other than Jupiter itself.

The "English/Lick" object, on the other hand, was almost certainly a comet of small perihelion distance that approached and receded from behind the Sun. Only the central regions of the coma would be observable against the daylight sky for a very brief period close to perihelion passage.

C/1927 X1 (Skjellerup–Maristany)

The Great Comet of 1927 was visible with the naked eye in broad daylight close to the Sun in mid-December (see Chapter 5 for an account of this comet).

C/1947 X1

The Great Southern Comet of 1947 was first seen as a brilliant object in the evening twilight on December 7 when it was described as possessing a strong orange color and a tail of 20–30 degrees length. Fading quickly, the display was brief, with the comet having dimmed below naked-eye visibility before Christmas. Perihelion had already occurred on December 2.59 at 0.11 AU and, according to the orbit, the comet should have been discovered earlier. It is interesting to note that the nucleus was found double on December 10, with both components remaining visible until the comet was lost in evening twilight on January 20, 1948. It has been shown that the split occurred near perihelion and was most probably responsible for a massive surge in brightness. The following account suggests that, initially, the surge may have been very large indeed!

Despite recorded magnitude estimates as bright as -5, I was aware of no daylight reports prior to receiving a telephone call in 1985 from Harold S. Pallot, who observed the comet from Horsham in Victoria on December 8, 1947. Pallot had been teaching astronomy classes at a local secondary school for several years prior to that year and was well known to the students of the school. At 6.30 in the afternoon (still daylight at that time of year) two teenagers (Graham Coster and Betty Hill) alerted him to the presence of a bright object in the western sky. Pallot saw the comet at 6.40 p.m. and, in a follow-up letter to me, writes that "It was not 'star' shiny, but rather 'flat' white with a very slight appearance of yellow in the coma. But the astonishing thing was, despite brilliant sunshine, the whole comet from end to end was clear and bright." He also described the comet as being much brighter than Venus at the time of the sighting.

C/1948 V1

The "Eclipse Comet" of 1948 approached from behind the Sun, passing perihelion unseen on October 27 at 0.14 AU. On November 1, during a total eclipse of the Sun observed from Nairobi, the comet suddenly became visible as a brilliant object some 105 min of arc from the Sun's center and showing a very strongly curved tail visible for at least 4 degrees on photographs. The comet was visible very briefly into the partial phase of the eclipse, but soon became lost against the brightening sky. It was rediscovered on November 4 by Captain Frank McGann from an aircraft and on the following mornings by numerous people as the bright long-tailed object emerged from the rays of the dawn. By mid-November, its tail extended some 30 degrees as observed with the naked eye. The comet faded below naked-eye visibility around December 20.

In the late 1980s, a short article in the journal of an Australian astronomical society recounts an experience of the writer who, late one afternoon at the end of October 1948, was sitting on the porch of his house watching the Sun set when his attention was caught by a bright comet with a tail next to it. He recalled that a dense layer of haze, plus the Sun's low altitude, made it possible to look in the direction of the Sun without being dazzled and it was because of this that he noticed the presence of the comet. Thinking that such a bright object would certainly be known to astronomers (!), he did not report his sighting – at least, not until some 40 years later!

Unfortunately, the details of this observation have become lost, but it seems to be genuine and probably took place on the evening of October 31.

C/1965 S1 (Ikeya–Seki)

One of the brightest comets on record, this object was clearly visible with the naked-eye in broad daylight near perihelion (see Chapter 6 for an account of this Kreutz sungrazer).

C/1973 E1 (Kohoutek)

This comet was discovered photographically on March 18, 1973, when only magnitude 16. When its orbit was calculated, it was found that perihelion (at 0.14 AU) would not occur until December 28, providing what was then an unprecedented time to prepare for a large comet. The comet was visible with the unaided eye from late November 1973 until late January 1974, with a faint tail of around 20 degrees length visible mid-January.

In 1975, during an address to the Japanese comet observing society *Hoshino Hiroba* John Bortle asked if anybody had attempted to observe this comet in daylight. He received a positive reply from amateur astronomer Kalsahito Mameta of Kobe, who saw it just after sunrise on December 27. By blocking the Sun from view behind the corner of a nearby house, he was able to find the comet with a pair of 16 x 50 binoculars a couple of degrees further west. Mameta estimated the head of the comet as several minutes of arc in size and described it as having "wings" flanking the central condensation and sweeping back away from the Sun. He kept the comet in view for several minutes and guessed the brightness as –6, almost certainly an overestimate. Mameta's drawing of the comet relative to the Sun agrees very well with its true position, further confirming the accuracy of his observation. Skylab astronauts saw the comet about the same time and estimated its brightness as between 0 and -3. In view of Mameta's observation, the brighter range of these estimates is likely correct (I am grateful to John Bortle for supplying the details of this little-known observation).

C/1975 V1 (West)

The Great Comet of 1976 became visible in daylight – both telescopically and with the naked eye – near perihelion (see Chapter 5 for an account of this comet).

C/1995 O1 (Hale–Bopp)

The Great Comet of 1997 was a marginal telescopic daylight object, but its daylight visibility extended over an unusually extensive period of time (see Chapter 5 for details of this comet).

C/1998 J1 (SOHO)

This comet approached from the far side of the Sun and remained hidden in twilight until May 3, when it suddenly emerged as a bright (magnitude 0) object in images by the LASCO C3 coronagraph on board the SOHO spacecraft. To date, this is the only comet discovered by LASCO that was later observed from the ground. It reached perihelion on May 8 at 0.17 AU and entered evening twilight on May 9, becoming visible with the naked eye by the middle of the month. By May 20, it displayed some 4 degrees of tail with the naked eye and fully 8 degrees in binoculars.

While visible in SOHO/LASCO images, several attempts were made to spot the comet in daylight, mostly by observers using reflectors equipped with digital setting circles. These were unsuccessful, however Mr. Fraser Farrell (Christies Beach, South Australia) swept the region near the Sun on May 8 using a 15-cm (6-in.) reflector and located a dim triangular object about one minute of arc in diameter, at least approximately in the region of the comet. This was kept under observation for 15 min, during which time it shared the diurnal motion typical of an astronomical body. Bright stars were also visible in Farrell's telescope, but had a different appearance to this object. A piece of wind-blown debris or a distant weather balloon seems unlikely in view of the diurnal motion and a tiny cloudlet is unlikely to have remained stable for the duration of the observation. The most likely explanation is a genuine daylight sighting of the comet. The lack of success experienced by other observers was explained once the comet became visible in the evening twilight. There was an unexpected distortion in the SOHO data which meant that early orbital computations (based *only* on this data) were slightly in error and the comet was not positioned where telescopes relying upon digital setting circles were pointing!

C/2006 P1 (McNaught)

The Great Comet of 2007 was the most spectacular comet of recent years.(see Chapter 5 for an account of this magnificent object).

It is fitting that our list of daylight comets ends with Comet McNaught, without doubt the most widely observed (though not the most brilliant) daylight comet yet to have graced the heavens. We cannot know when its equal will arrive, or when there will be another comet visible while the Sun is above the horizon, but there can be little doubt that this young century holds many cometary surprises in store for future years. I just hope that, for the sake of the new generation of comet enthusiasts, it proves as rich in truly great comets as the last two centuries were!

Glossary

Absolute magnitude For Solar System objects, the magnitude of an object located at one Astronomical Unit from both Earth and Sun.

Aphelion The point of an orbit farthest from the Sun.

Asteroids (Minor Planets) Predominantly rocky bodies smaller than the major planets and found mostly within the inner Solar System.

Astronomical Unit (AU) A unit of distance equal to the average distance of Earth from the Sun, predominantly used for expressing distances within the Solar System. One AU is approximately 92,955.807 miles, or 149,597,870 km.

Averted vision The technique of looking slightly away from a faint object so as to bring its image to the most light-sensitive part of the retina.

Celestial poles The points in the heavens directly over the north or south poles of Earth.

Coma The cloud of dust and gas that forms around the nucleus of a comet when the latter is heated by the Sun. The nucleus and coma together make the head of a comet.

Comet A predominantly icy body capable of producing clouds of dust and gas when heated by the Sun.

Degree Angular measure in which a circle around the entire heavens above and below the horizon can be divided into 360 degrees of arc; 1 degree can be divided into 60 min of arc and 1 min can be divided into 60 s of arc.

Dynamically new comet A comet that is apparently entering the inner Solar System from the Oort Cloud for the first time.

Eccentricity An orbital element that defines how elongated or even how open-ended the orbit of an astronomical object is. A circle has an eccentricity of 0, an ellipse between 0 and 1, a parabola exactly 1, and a hyperbola greater than 1.

Ecliptic plane The plane of Earth's orbit around the Sun.

Great comet A comet that becomes unusually spectacular and conspicuous with the naked eye.

Head (of a comet) A name given to the combined nucleus and coma of a comet, especially when a visible tail is also present.

Inclination An orbital element that defines the angle between the plane of an object's orbit and the plane of Earth's orbit (i.e., the ecliptic plane).

Kuiper Belt A system of comets and icy dwarf planets (including Pluto) just beyond the orbits of the outermost planets in the Solar System. This is believed to be the source of comets having short periods.

Magnitude A measure of brightness, used in astronomy, in which a difference of 5 magnitudes equals a difference in brightness of 100 times. The lower the magnitude, the brighter the object.

Meteor The streak of light in the sky caused by a meteoroid burning up in Earth's atmosphere.

Meteorite An interplanetary piece of rock or iron reaching the surface of a moon or planet, especially Earth.

Meteoroid A small solid object in orbit around the Sun which, upon entry into Earth's atmosphere, gives rise to a meteor.

Nucleus The solid, more permanent part of a comet and the source of its coma and tail.

Oort Cloud The vast "cloud" of comets, beyond 50,000 AU from the Sun, believed to be the source of comets having very long periods.

Orbital Elements A set of quantities that specifies the size and shape of an orbit and the times when the object following that orbit reaches key positions.

Parallax The apparent shift of a nearby object's position in relation to more distant ones, when the former is viewed from different viewing angles.

Perihelion The point of an orbit nearest to the Sun.

Plasma A gas composed of electrons and positive ions.

Return The reappearance of a comet at a specific perihelion passage.

Solar radiation pressure The force exerted by photons of visible light and other wavelengths of electromagnetic radiation.

Solar wind The plasma constantly traveling outward from the Sun.

Swan bands Three prominent bands in the spectrum of most comets, caused by diatomic carbon (C_2).

Synchrone A line connecting dust particles in a comet tail that left the nucleus at the same time.

Syndyne (syndyname) A line connecting particles in a comet tail that left the nucleus at the same velocity.

Tail An appendage of gas and/or dust that streams away from the coma of many comets under the influence of solar radiation pressure or the solar wind.

Further Reading

Most books on general astronomy have a chapter dedicated to comets. Of the books published from ca. 1980 and dealing specifically with comets, the following are pleasantly non-technical.

Burnham, Robert. *Great Comets*. Cambridge: Cambridge University Press, 2000

Kronk, Gary W. *Cometography* (in six volumes) Cambridge: Cambridge University Press. First three volumes published 1999, 2003, and 2007.

Kronk, Gary W. *Comets: A Descriptive Catalog*. Hillside, NJ: Enslow Publishers Inc., 1984.

Levy, David H. *The Quest for Comets*. New York: Avon Books, 1995.

Levy, David. *Comets: Creators and Destroyers*. New York: Touchstone Books, 1998.

Schaaf, Fred. *Comet of the Century: From Halley to Hale-Bopp*. Springer-Verlag, New York, Inc., 1997.

Appendix

The following table lists all comets that we consider to have been the "greatest" of the many objects recorded throughout history. Listed are the official designation (where applicable) or year of appearance where no official designation has been given, the date of perihelion (T), perihelion distance in AU (q), minimum distance from earth (Δmin), the date of minimum approach to Earth (DateΔmin), approximate apparent magnitude at greatest observed brilliance (m), approximate visual length of tail, in degrees, at maximum (Tail) and approximate absolute magnitude ($H10$). The last three are, in most instances, little more than guesses – "wild" guesses for the early entries – but will hopefully facilitate some degree of comparison between these magnificent comets.

Comet	T	q	Δmin	DateΔmin	m	Tail	H10
372 B.C	Winter	v small			v bright	60	
135 B.C (−134 N1)					v bright	>90?	
44 B.C (−43 K1)	May 25	0.22	0.96	May 12	−3?	15	3
P/141 F1 (1P/Halley)	March 22.43	0.58	0.17	April 21	−1	>10	3
178	Early September	0.5	0.05	Mid-October	Bright	90?	
191	Early September?	<0.01?				>100	
252	March 13–21	<0.01				90?	
P/374 E1 (1P/Halley)	February 16.34	0.58	0.09	April 1	−2	90?	3
C/390 Q1	September 5	0.92	0.1	August 18	−3	100	
C/400 F1	February 25	0.21	0.07	March 31	0	45	6
C/418 M1	October 5	0.35	0.93	October 9	−4 ?	>100	−2?
467	January?	<0.01?				Long	
P/607 U1 (1P/Halley)	March 15.47	0.58	0.09	April 19	−2	90?	3
X/676 P1						45	
C/770 K1	June 5.8	0.58	0.30	July 10	−3	75	3.2
P/837 F1 (1P/Halley)	February 28.27	0.58	0.03	April 10	−4	90	3
X/838 V1						100	
X/891 J1						>100	
893					Very brilliant	>100(200?)	
C/905 K1	April 26	0.20	0.20	May 25	0	>100	4.5
P/1066 G1	March 20.93	0.57	0.10	April 23	−5?	>100	3

(continued)

Comet	T	q	Δmin	DateΔmin	m	Tail	H10
(1P/Halley)							
X/1106 C1	Early February?	<0.01?			−10?	100	
C/1132 T1	August 30.7	0.74	0.05	October 6	−2	45	4.5
C/1147 A1	January 28	0.12	0.32	December 29 (1146)	0	100	5
C/1264 N1	July 20.29	0.82	0.18	July 29	0	100	3.5
C/1402 D1	March 21	0.38	0.71	February 20	−5	45	0
C/1471 Y1	March 1.44(1472)	0.49	0.07	January 22	−4	30	2
C/1577 V1	October 27.45	0.18	0.63	November 10	−4	20	0
C/1582 J1	May 6.91	0.17	0.83	May 8	−1	100	4.8
Gregorian calendar							
C/1618 W1	November 8.85	0.39	0.36	December 6	−4?	104	4.6
C/1680 V1	November 18.49	0.006	0.49	January 4 (1681)	−10	90	4
C/1743 X1	March 1.89(1744)	0.22	0.83	February 26 (1744)	−7	90	0.5
C/1769 P1	October 8.12	0.12	0.32	September 10	0	90	3.2
C/1811 F1	September 12.76	1.04	1.22	October 16	0	25	0
C/1843 D1	February 27.91	0.006	0.84	March 5	−11	64	4.9
C/1858 L1	September 30.47	0.58	0.54	October 10	−1	43	3.3
C/1861 J1	June 12.01	0.82	0.13	June 30	−5	122	3.9
C/1882 R1	September 17.72	0.008	0.98	September 17	−12.5	30	0.8
P/1909 R1	April 20.18(1910)	0.58	0.15	May 20 (1910)	−6?	>120	3
(P/Halley)							
C/1910 A1	January 17.59	0.13	0.86	January 18	−6	50	5
C/1927 X1	December 18.18	0.18	0.75	December 12	−9	35	5.5
C/1965 S1	October 21.18	0.008	0.91	October 17	−15	45	6
C/1975 V1	February 25.22(1976)	0.20	0.79	March 1 (1976)	−3	40	4.5
C/1995 O1	March 30.29(1997)	0.93	1.31	March 22 (1997)	−0.3	40	0
C/1996 B2	May 1.40	0.23	0.10	March 22	0	118	5
C/2006 P1	January 12.80 (2007)	0.17	0.82	January 15 (2007)	−6	40	5.5

Name Index

Subject Index

Printed in the United States